Large-Scale Computing

**WILEY SERIES ON PARALLEL
AND DISTRIBUTED COMPUTING**

Editor: Albert Y. Zomaya

A complete list of titles in this series appears at the end of this volume.

Large-Scale Computing

Edited by

Werner Dubitzky
University of Ulster, United Kingdom

Krzysztof Kurowski
Poznan Supercomputing and Networking Center, Poland

Bernhard Schott
Platform Computing, GmbH, Germany

A John Wiley & Sons, Inc., Publication

Library of Congress Cataloging-in-Publication Data:

Large-scale computing techniques for complex system simulations / [edited by] Werner
Dubitzky, Krzysztof Kurowski, Bernhard Schott.
    p. cm.—(Wiley series on parallel and distributed computing ; 80)
  Includes bibliographical references and index.
  ISBN 978-0-470-59244-1 (hardback)
  1. Computer simulation. I. Dubitzky, Werner, 1958– II. Kurowski, Krzysztof,
1977– III. Schott, Bernhard, 1962–
  QA76.9.C65L37 2012
  003'.3–dc23

                                                                        2011016571

eISBN 9781118130476
oISBN 9781118130506
ePub 9781118130490
MOBI 9781118130483

Printed in United States of America.

10 9 8 7 6 5 4 3 2 1

Contents

Foreword

The human desire to understand things is insatiable and perpetually drives the advance of society, industry, and our quality of living. Scientific inquiry and research are at the core of this desire to understand, and, over the years, scientists have developed a number of formal means of doing that research. The third means of scientific research, developed in the past 100 years after theory and experimentation, is modeling and simulation. The sharpest tool invented thus far for modeling and simulation has been the computer. It is fascinating to observe that, within the short span of a few decades, humans have developed dramatically larger and larger scale computing systems, increasingly based on massive replication of commodity elements to model and simulate complex phenomena in increasing details, thus delivering greater insights. At this juncture in computing, it looks as if we have managed to turn everything physical into its virtual equivalent, letting a virtual world precede the physical one it reflects so that we can predict what is going to happen before it happens. Better yet, we want to use modeling and simulation to tell us how to change what is going to happen. The thought is almost scary as we are not supposed to do God's work.

We all know the early studies of ballistic trajectories and code breaking, which stimulated the development of the first computers. From those beginnings, all kinds of physical objects and natural phenomena have been captured and reflected in computer models, ranging from particles in a "simple" atom to the creation of the universe, with modeling the earth's climate in between.

Computer-based modeling systems are not just limited to natural systems, but increasingly, man-made objects are being modeled, as well. One could say that, without computers, there would not have been the modern information, communication, and entertainment industries because the heart of these industries' products, the electronic chips, must be extensively simulated and tested before they are manufactured.

Even much larger physical products, like cars and airplanes, are also modeled by computers before they go into manufacturing. An airbus virtually assembles all of its planes' graphically rendered parts every night to make sure they fit together and work, just like a software development project has all its pieces of code compiled and regression tested every night so that the developers can get reports on what they did wrong the day before to fix their problems. Lately, modelers have advanced to simulating even the human body itself, as well as the organizations we create: How do drugs interact with the proteins in our bodies? How can a business operate more efficiently to generate more revenue and profits by optimizing its business processes? The hottest area of enterprise computing applications nowadays is business analytics.

Tough problems require sharp tools, and imaginative models require innovative computing systems. In computing, the rallying cry from the beginning has been "larger is better": larger computing power, larger memory, larger storage, larger everything. Our desire for more computing capacity has been insatiable. To solve the problems that computer simulations address, computer scientists must be both dreamers and greedy for more computing power at the same time. And it is all for the better: It is valuable to be able to simulate in detail a car with a pregnant woman sitting in it and to model how the side and front air bags will function when surrounded by all the car parts and involved in all kinds of abnormal turns. More accurate results require using more computing power, more data storage space, and more interconnect bandwidth to link them all together. In computing, greed is good—if you can afford it.

That is where this book will start: How can we construct infinitely powerful computing systems with just the right applications to support modeling and simulating problems at a very low cost so they can be accessible to everybody? There is no simple answer, but there are a million attempts. Of course, not all of them lead to real progress, and not all of them can be described between the two covers of a book. This book carefully selected eight projects and enlisted their thinkers to show and tell what they did and what they learned from those experiences. The reality of computing is that it is still just as much an art as it is science: it is the cleverness of how to put together the silicon and the software programs to come up with a system that has the lowest cost while also being the most powerful and easiest to use. These thinkers, like generations of inventors before them, are smart, creative people on a mission.

There have certainly been many innovations in computer simulation systems, but three in particular stand out: commodity replication, virtualization, and cloud computing. Each of these will also be explored in this book, although none of these trends have been unique to simulation.

Before commodity replication, the computing industry had a good run of complex proprietary systems, such as the vector supercomputers. But when it evolved to 30 layers in a printed circuit board and a cost that could break a regional bank, it had gone too far. After that, the power was in the replication of commodity components like the x86 chips used in PCs, employing a million of them while driving a big volume discount, then voilà, you get a large-scale system at a low(er) cost! That is what drove computing clusters and grids.

The complexity animal was then tamed via software through virtualization, which abstracts away the low-level details of the system components to present a general systems environment supporting a wide array of parallel programming environments and specialized simulations. Virtualization allows innovation in computer system components without changing applications to fit the systems. Finally, cloud computing may allow users not to have to actually own large simulation computers anymore but rather to only use computing resources as needed, sharing such a system with other users. Or even better, cloud resources can be rented from a service provider and users only pay for usage. That is an innovation all right, but it feels like we are back to the mainframe service bureau days. In this book, you will learn about many variations of innovations around these three themes as applied to simulation systems.

So, welcome to a fascinating journey into the virtual world of simulation and computing. Be prepared to be perplexed before possibly being enlightened. After all, simulation is supposed to be the state of the art when it comes to complex and large-scale computing. As they say, the rewards make the journey worthwhile.

Songnian Zhou
Toronto, Canada
March 2011

Preface

Complex systems are defined as systems with many interdependent parts that give rise to nonlinear and emergent properties determining their high-level functioning and behavior. Due to the interdependence of their constituent elements and other characteristics of complex systems, it is difficult to predict system behavior based on the "sum of their parts" alone. Examples of complex systems include human economies and societies, nervous systems, molecular interaction networks, cells and other living things, such as bees and their hives, and ecosystems, as well as modern energy and telecommunication infrastructures. Arguably, one of the most striking properties of complex systems is that conventional experimental and engineering approaches are inadequate to capture and predict the behavior of such systems. A relatively recent and more holistic approach employs computational techniques to model and simulate complex natural phenomena and complex man-made artifacts. Complex system simulations typically require considerable computing and storage resources (processing units and primary and secondary memory) as well as high-speed communication links. Supercomputers are the technology of choice to satisfy these requirements. Because supercomputers are expensive to acquire and maintain, there has been a trend to exploit distributed computing and other large-scale computing technologies to facilitate complex system simulations. Grid computing, service-oriented architectures, programmable logic arrays, and graphic processors are examples of such technologies.

The purpose of this volume is to present a representative overview of contemporary large-scale computing technologies in the context of complex system simulation applications. The book is intended to serve simultaneously as design blueprint, user guide, research agenda, and communication platform. As a *design blueprint*, the book is intended for researchers and technology and application developers, managers, and other professionals who are tasked with the development or deployment of large-scale computer technology to facilitate complex system applications. As a *user guide*, the volume addresses the requirements of modelers and scientists to gain an overview and a basic understanding of key concepts, methodologies, technologies, and tools. For this audience, we seek to explain the key concepts and assumptions of the various large-scale computer techniques and technologies, their conceptual and computational merits and limitations. We aim at providing the users with a clear understanding and practical know-how of the relevant technologies in the context of complex system modeling and simulation and the large-scale computing technologies employed to meet the requirements of such applications. As *research agenda*, the book is intended for computer and complex systems students, teachers, and researchers who seek to understand the state of the art of the large-scale computing technologies involved as well as their limitations and emerging and future developments. As a *communication platform*, the book is intended to bridge the cultural, conceptual, and technological gap among the key disciplines of complex system modeling and simulation and large-scale computing. To support this goal, we have asked the contributors to adopt an approach that appeals to audiences from different backgrounds.

Clearly, we cannot expect to do full justice to all of these goals in a single book. However, we do believe that this book has the potential to go a long way in fostering the understanding, development, and deployment of large-scale computer technology and its application to the modeling and simulation of complex systems. Thus, we hope this volume will contribute to increased communication and collaboration across various modeling, simulation, and computer science disciplines and will help to improve the complex natural and engineering systems.

This volume comprises nine chapters, which introduce the key concepts and challenges and the lessons learned from developing and deploying large-scale computing technologies in the context of complex system applications. Next, we briefly summarize the contents of the nine chapters.

Chapter 1 is concerned with an overview of some large-scale computing technologies. It discusses how in the last three decades the demand for computer-aided simulation of processes and systems has increased. In the same time period, simulation models have become increasingly complex in order to capture the details of the systems and processes being modeled. Both trends have instigated the development of new concepts aimed at a more efficient sharing of computational resources. Typically, grid and cloud computing techniques are employed to meet the computing and storage demands of complex applications in research, development, and other areas. This chapter provides an overview of grid and cloud computing, which are key elements of many modern large-scale computing environments.

Chapter 2 adopts the view of an e-infrastructure ecosystem. It focuses on scientific collaborations and how these are increasingly relying on the capability of combining computational and data resources supplied by several resource providers into seamless e-infrastructures. This chapter presents the rationale for building an e-infrastructure ecosystem that comprises national, regional, and international e-infrastructures. It discusses operational and usage models and highlights how e-infrastructures can be used in building complex applications.

Chapter 3 presents multiscale physical and astrophysical simulations on new many-core accelerator hardware. The chosen algorithms are deployed on parallel clusters using a large number of graphical processing units (GPUs) on the petaflop scale. The applications are particle-based astrophysical many-body simulations with self-gravity, as well as particle and mesh-based simulations on fluid flows, from astrophysics and physics. Strong and soft scaling are demonstrated using some of the fastest GPU clusters in China and hardware resources of cooperating teams in Germany and the United States.

Chapter 4 presents an overview of the SimWorld Agent-Based Grid Experimentation System (SWAGES). SWAGES has been used extensively for various kinds of agent-based modeling and is designed to scale to very large and complex grid environments while maintaining a very simple user interface for integrating models with the system. This chapter focuses on SWAGES' unique features for parallel simulation experimentation (such as novel spatial scheduling algorithms) and on its methodologies for utilizing large-scale computational resources (such as the distributed server architecture designed to offset the ever-growing computational demands of administering large simulation experiments).

Chapter 5 revolves around agent-based modeling and simulation (ABMS) technologies. In the last decade, ABMS has been successfully applied to a variety of domains, demonstrating the potential of this approach to advance science, engineering, and other domains. However, realizing the full potential of ABMS to generate breakthrough research results requires far greater computing capability than is available through current ABMS tools. The Repast for High Performance Computing (Repast HPC) project addresses this need by developing a next-generation ABMS system explicitly focusing on larger-scale distributed computing platforms. This chapter's contribution is its detailed presentation of the implementation of Repast HPC, a complete ABMS platform developed explicitly for large-scale distributed computing systems.

Chapter 6 presents an environment for the development and execution of multiscale simulations composed from high-level architecture (HLA) components. Advanced HLA mechanisms are particularly useful for multiscale simulations as they provide, among others, time management functions that enable the construction of integrated simulations from modules with different individual timescales. Using the proposed solution simplifies the use of HLA services and allows components to be steered by users; this is not possible in raw HLA. This solution separates the roles of simulation module developers from those of users and enables collaborative work. The environment is accessible via a scripting API, which enables the steering of distributed components using straightforward source code.

Chapter 7 is concerned with the data dimensions of large-scale computing. Data-intensive computing is the study of the tools and techniques required to manage and explore digital data. This chapter briefly discusses the many issues arising from the huge increase in stored digital data that we are now confronted with globally. In order to make sense of this data and to transform it into useful information that can inform our knowledge of the world around us, many new techniques in data handling, data exploration, and information creation are needed. The Advanced Data Mining and Integration Research for Europe (ADMIRE) project, which this chapter discusses in some detail, is studying how some of these challenges can be addressed through the creation of advanced, automated data mining techniques that can be applied to large-scale distributed data sets.

Chapter 8 describes a topology-aware evolutionary algorithm that is able to automatically adapt itself to different configurations of distributed computing resources. An important component of the algorithm is the use of QosCosGrid-OpenMPI, which enables the algorithm to run across computing resources hundreds of kilometers distant from one another. The authors use the evolutionary algorithm to compare models of a biological gene regulatory network which have been reverse engineered using three different systems of mathematical equations.

Chapter 9 presents a number of technologies that have been successfully integrated into a supercomputing-like e-science infrastructure called QosCosGrid (QCG). The solutions provide services for simulations such as complex systems, multiphysics, hybrid models, and parallel applications. The key aim in providing these solutions was to support the dynamic and guaranteed use of distributed computational clusters and supercomputers managed efficiently by a hierarchical scheduling structure involving a metascheduler layer and underlying local queuing or batch systems.

Coleraine, Frankfurt, Poznan   Werner Dubitzky
May 2011   Krzysztof Kurowski
                                                   Bernhard Schott

Contributors

PETER BERCZIK, National Astronomical Observatories of China, Chinese Academy of Sciences, China, and Astronomisches Rechen-Institut, Zentrum für Astronomie, University of Heidelberg, Germany; Email: berczik@bao. ac.cn

INGO BERENTZEN, Zentrum für Astronomie, University of Heidelberg, Heidelberg, Germany; Email: iberent@ari.uni-heidelberg.de

BARTOSZ BOSAK, Poznan Supercomputing and Networking Center, Poznan, Poland; Email: bbosak@man.poznan.pl

MARIAN BUBAK, AGH University of Science and Technology, Krakow, Poland, and Institute for Informatics, University of Amsterdam, The Netherlands; Email: bubak@agh.edu.pl

FRANCK CAPPELLO, National Institute for Research in Computer Science and Control (INRIA), Rennes, France; Email: franck.cappello@inria.fr

NICHOLSON COLLIER, Argonne National Laboratory, Argonne IL; Email: ncollier@anl.gov

CAMILLE COTI, LIPN, CNRS-UMR7030, Université Paris 13, F-93430 Villetaneuse, France; Email: camille.coti@lipn.univ-paris13.fr

WERNER DUBITZKY, School of Biomedical Sciences, University of Ulster, Coleraine BT52 1SA, UK; Email: w.dubitzky@ulster.ac.uk

ÅKE EDLUND, KTH Royal Institute of Technology, Stockholm, Sweden; Email: edlund@nada.kth.se

CHAKER EL AMRANI, Université Abdelmalek Essaâdi, Tanger, Morocco; Email: ch.elamrani@fstt.ac.ma

FLORIAN FELDHAUS, Dortmund University of Technology, Dortmund, Germany; Email: florian.feldhaus@udo.edu

JOSÉ FIESTAS, Astronomisches Rechen-Institut, Zentrum für Astronomie, University of Heidelberg, Germany; Email: fiestas@ari.uni-heidelberg.de

STEFAN FREITAG, Dortmund University of Technology, Dortmund, Germany; Email: stefan.freitag@udo.edu

WEI GE, Institute of Process Engineering, Chinese Academy of Sciences, Beijing, China; Email: wge@home.ipe.ac.cn

PIOTR GRABOWSKI, Poznan Supercomputing and Networking Center, Poznan, Poland; Email: piotrg@man.poznan.pl

LÁSZLÓ GULYÁS, Aitia International Inc. and Collegium Budapest (Institute for Advanced Study), Budapest, Hungary; Email: gulya@aitia.ai

TSUYOSHI HAMADA, Nagasaki Advanced Computing Center, Nagasaki University, Nagasaki, Japan; Email: hamada@nacc.nagasaki-u.ac.jp

JACK J. HARRIS, Human Robot Interaction Laboratory, Indiana University, Bloomington, IN; Email: jackharr@indiana.edu

THOMAS HERAULT, National Institute for Research in Computer Science and Control (INRIA), Rennes, France; Email: thomas.herault@lri.fr

GEORGE KAMPIS, Collegium Budapest (Institute for Advanced Study), Budapest, Hungary; Email: kampis.george@gmail.com

KRZYSZTOF KUROWSKI, Poznan Supercomputing and Networking Center, Poznan, Poland; Email: krzysztof.kurowski@man.poznan.pl

ERWIN LAURE, KTH Royal Institute of Technology, Stockholm, Sweden; Email: erwinl@pdc.kth.se

MARIUSZ MAMONSKI, Poznan Supercomputing and Networking Center, Poznan, Poland; Email: mmamonski@man.poznan.pl

JOHANNES MANDEL, Roche Diagnostics GmbH, Penzberg, Germany; Email: johannes.mandel@roche.com

KEIGO NITADORI, RIKEN AICS Institute, Kobe, Japan; Email: keigo@riken.jp

MICHAEL NORTH, Argonne National Laboratory, Argonne IL; Email: north@anl.gov

MARK PARSONS, EPCC, The University of Edinburgh, Edinburgh, UK; Email: m.parsons@epcc.ed.ac.uk

TOMASZ PIONTEK, Poznan Supercomputing and Networking Center, Poznan, Poland; Email: piontek@man.poznan.pl

KATARZYNA RYCERZ, AGH University of Science and Technology, Krakow, Poland, and ACC Cyfronet AGH, Krakow, Poland; Email: kzajac@agh.edu.pl

MATTHIAS SCHEUTZ, Department of Computer Science, Tufts University, Medford, MA; Email: mscheutz@cs.tufts.edu

HSI-YU SCHIVE, Department of Physics, National Taiwan University, Taibei, Taiwan; Email: b88202011@ntu.edu.tw

RAINER SPURZEM,   National Astronomical Observatories of China, Chinese Academy of Sciences, China; Astronomisches Rechen-Institut, Zentrum für Astronomie, University of Heidelberg, Germany; and Kavli Institute for Astronomy and Astrophysics, Peking University, China; Email: spurzem@ bao.ac.cn

MARTIN SWAIN,   Institute of Biological, Environmental and Rural Sciences, Aberystwyth University, Penglais, Aberystwyth, Ceredigion, UK; Email: mts11@aber.ac.uk

XIAOWEI WANG,   Institute of Process Engineering, Chinese Academy of Sciences, Beijing, China; Email: xwwang@home.ipe.ac.cn

Chapter *1*

State-of-the-Art Technologies for Large-Scale Computing

Florian Feldhaus and Stefan Freitag

Dortmund University of Technology, Dortmund, Germany

Chaker El Amrani

Université Abdelmalek Essaâdi, Tanger, Morocco

1.1 INTRODUCTION

Within the past few years, the number and complexity of computer-aided simulations in science and engineering have seen a considerable increase. This increase is not limited to academia as companies and businesses are adding modeling and simulation to their repertoire of tools and techniques. Computer-based simulations often require considerable computing and storage resources. Initial approaches to address the growing demand for computing power were realized with supercomputers in 60 seconds. Around 1964, the CDC6600 (a mainframe computer from Control Data Corporation) became available and offered a peak performance of approximately 3×10^6 floating point operations per second (flops) (Thornton, 1965). In 2008, the IBM Roadrunner[1] system, which offers a peak performance of more than 10^{15} flops, was commissioned into service. This system was leading the TOP500 list of supercomputers[2] until November 2009.

[1] www.lanl.gov/roadrunner.
[2] www.top500.org.

Large-Scale Computing, First Edition. Edited by Werner Dubitzky, Krzysztof Kurowski, Bernhard Schott.
© 2012 John Wiley & Sons, Inc. Published 2012 by John Wiley & Sons, Inc.

Supercomputers are still utilized to execute complex simulations in a reasonable amount of time, but can no longer satisfy the fast-growing demand for computational resources in many areas. One reason why the number of available supercomputers does not scale proportional to the demand is the high cost of acquisition (e.g., $133 million for Roadrunner) and maintenance.

As conventional computing hardware is becoming more powerful (processing power and storage capacity) and affordable, researchers and institutions that cannot afford supercomputers are increasingly harnessing *computer clusters* to address their computing needs. Even when a supercomputer is available, the operation of a local cluster is still attractive, as many workloads may be redirected to the local cluster and only jobs with special requirements that outstrip the local resources are scheduled to be executed on the supercomputer.

In addition to current demand, the acquisition of a cluster computer for processing or storage needs to factor in potential increases in future demands over the computer's lifetime. As a result, a cluster typically operates below its maximum capacity for most of the time. E-shops (e.g., Amazon) are normally based on a computing infrastructure that is designed to cope with peak workloads that are rarely reached (e.g., at Christmas time).

Resource providers in academia and commerce have started to offer access to their underutilized resources in an attempt to make better use of spare capacity. To enable this provision of free capacity to third parties, both kinds of provider require technologies to allow remote users restricted access to their local resources. Commonly employed technologies used to address this task are *grid computing* and *cloud computing*. The concept of grid computing originated from academic research in the 1990s (Foster et al., 2001). In a grid, multiple resources from different administrative domains are pooled in a shared infrastructure or computing environment. Cloud computing emerged from commercial providers and is focused on providing easy access to resources owned by a single provider (Vaquero et al., 2009).

Section 1.2 provides an overview of grid computing and the architecture of grid middleware currently in use. After discussing the advantages and drawbacks of grid computing, the concept of virtualization is briefly introduced. Virtualization is a key concept behind cloud computing, which is described in detail in Section 1.4. Section 1.5 discusses the future and emerging synthesis of grid and cloud computing before Section 1.7 summarizes this chapter and provides some concluding remarks.

1.2 GRID COMPUTING

Foster (2002) proposes three characteristics of a grid:

1. Delivery of nontrivial qualities of service
2. Usage of standard, open, general-purpose protocols and interfaces
3. Coordination of resources that are not subject to centralized control

Endeavors to implement solutions addressing the concept of grid computing ended up in the development of grid middleware. This development was and still is driven by communities with very high demands for computing power and storage capacity. In the following, the main grid middleware concepts are introduced and their implementation is illustrated on the basis of the gLite[3] middleware, which is used by many high-energy physics research institutes (e.g., CERN). Other popular grid middleware include Advanced Resource Connector (ARC[4]), Globus Toolkit,[5] National Research Grid Initiative (NAREGI[6]), and Platform LSF MultiCluster.[7]

Virtual Organizations. A central concept of many grid infrastructures is a *virtual organization*. The notion of a virtual organization was first mentioned by Mowshowitz (1997) and was elaborated by Foster et al. (2001) as "a set of the individuals and/or institutions defined by resource sharing rules."

Virtual organization is used to overcome the temporal and spatial limits of conventional organizations. The resources shared by a virtual organization are allowed to change dynamically: Each participating resource/institution is free to enter or leave the virtual organization at any point in time. One or more resource providers can build a grid infrastructure by using grid middleware to offer computing and storage resources to multiple virtual organizations. Resources at the same location (e.g., at an institute or computing center) are forming a (local) grid site. Each grid site offers its resources through grid middleware services to the grid. For the management and monitoring of the grid sites as well as the virtual organizations, central services are required. The main types of service of a grid middleware may be categorized into (Fig. 1.1) (Foster, 2005; Burke et al., 2009) the following:

- Execution management
- Data management
- Information services
- Security

Execution Management. The execution management services deal with monitoring and controlling compute tasks. Users submit their compute tasks together with a description of the task requirements to a central workload management system (WMS). The WMS schedules the tasks according to their requirements to free resources discovered by the information system. As there may be thousands of concurrent tasks to be scheduled by the WMS, sophisticated scheduling mechanisms are needed. The simulation and analysis of the

---

[3] http://glite.cern.ch/.
[4] www.nordugrid.org/middleware.
[5] www.globus.org/toolkit.
[6] www.naregi.org/link/index_e.html.
[7] www.platform.com.

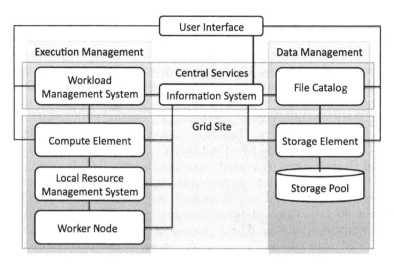

Figure 1.1   *Overview of grid middleware components.*

corresponding scheduling algorithms has become an important research area in its own right (Section 1.6).

Each grid site needs to run a compute element that is responsible for user authentication at the grid site and to act as an interface between local resources and the grid. The compute element receives compute tasks from the WMS and submits them to a local resource management system,[8] which then schedules the tasks to be executed on a free worker node.

With the LCG[9] and the CREAM[10] compute elements (Aiftimiei et al., 2010), the gLite middleware currently offers two choices for this task. Whereas the LCG is the standard compute element of gLite, the CREAM compute element was developed to be more lightweight and also allows direct job submission if mechanisms other than the central gLite WMS should be used for job scheduling and resource matching.

Data Management. Besides offering potentially powerful computing resources, a grid may also provide storage capacity in the form of storage elements.

With the Disk Pool Manager (DPM) (Abadie et al., 2007), the CERN Advanced STORage manager (CASTOR)[11] and dCache,[12] gLite currently supports three storage services. All of these are able to manage petabytes of storage on disk and/or tape. Retrieving and storing data are possible via

---

[8] Examples for resource management systems include TORQUE or the SUN Grid Engine.
[9] Large Hadron Collider Computing Grid.
[10] Computing Resource Execution and Management.
[11] http://castor.web.cern.ch/.
[12] www.dcache.org.

various protocols, for example, dcap, xrootd,[13] gridFTP, and SRM (Badino et al., 2009). The LCG storage element defines the minimum set of protocols that have to be supported to access the storage services.

For gLite, the central LCG File Catalog enables virtual organizations to create a uniform name space for data and to hide the physical data location. This is achieved by using logical file names that are linked to one or more physical file names (consisting of the full qualified name of the storage element and the absolute data path for the file on this specific storage element). Replicas of the data can be created by copying the physical files to multiple storage elements and by registering them under one unique logical file name in the LCG File Catalog. Thus, the risk of data loss can be reduced.

Information System. The information system discovers and monitors resources in the grid. Often the information system is organized hierarchically. The information about local resources is gathered by a service at each grid site and then sent to a central service. The central service keeps track of the status of all grid services and offers an interface to perform queries. The WMS can query the information to match resources to compute tasks.

For gLite, a system based on the Lightweight Directory Access Protocol, is used. The Berkeley Database Information Index (BDII) service runs at every site (siteBDII) and queries all local gLite services in a given time interval. In the same way, the siteBDII is queried by a TopLevelBDII, which stores the information of multiple grid resources. The list of available resources is kept in the information supermarket and is updated periodically by the TopBDII.

To account for the use of resources, a grid middleware may offer accounting services. Those register the amount of computation or storage used by individual or groups of users or virtual organizations. This allows billing mechanisms to be implemented and also enables the execution management system to schedule resources according to use history or quota.

The gLite Monitoring System Collector Server service gathers accounting data on the local resources and publishes these at a central service.

Security. To restrict the access of resources to certain groups of users or virtual organizations, authentication and authorization rules need to be enforced. To facilitate this, each virtual organization issues X.509 certificates to its users (user certificate) and resources (host certificate). Using the certificates, users and resources can be authenticated.

To allow services to operate on behalf of a user, it is possible to create proxy certificates. Those are created by the user and are signed with his user certificate. The proxy certificate usually only has a short lifetime (e.g., 24 hours) for security reasons. A service can the use the proxy certificate to authenticate against some other service on behalf of the user.

Authorization is granted according to membership in a virtual organization, a virtual organization group, or to single users. Access rights are managed centrally by the virtual organization for its users and resources.

[13] http://xrootd.slac.stanford.edu/.

To manage authorization information, gLite offers the Virtual Organization Management Service, which stores information on roles and privileges of users within a virtual organization. With the information from the Virtual Organization Management Service, a grid service can determine the access rights of individual users.

1.2.1 Drawbacks in Grid Computing

A grid infrastructure simplifies the handling of distributed, heterogeneous resources as well as users and virtual organizations. But despite some effort in this direction, the use of inhomogeneous software stacks (e.g., operating system, compiler, libraries) on different resources has been a weakness of grid systems.

Virtualization technology (see Section 1.4) offers solutions for this problem through an abstraction layer above the physical resources. This allows users, for example, to submit customized virtual machines containing their individual software stack. As grid middleware is not designed with virtualization in mind, the adaptation and adoption of virtualization technology into grid technology is progressing slowly. In contrast to grid computing, cloud computing (Section 1.5) was developed based on virtualization technology. Approaches to combine cloud and grid computing are presented in Section 1.6.

1.3 VIRTUALIZATION

As described in Section 1.2, grid computing is concerned mainly with secure sharing of resources in dynamic computing environments. Within the past few years, virtualization emerged and soon became a key technology, giving a new meaning to the concept of resource sharing.

This section describes two different types of technology (resource and platform virtualization) and provides an overview of their respective benefits.

Network Virtualization. At the level of the network layer, virtualization is categorized as *internal* and *external* virtualizations. External virtualization joins network partitions to a single virtual unit. Examples for this type of virtualization are virtual private networks and virtual local area networks (IEEE Computer Society, 2006). Internal virtualization offers network-like functionality to software containers on a resource. This type of virtualization is often used in combination with virtual machines.

Storage Virtualization. In mass storage systems, the capacities of physical storage devices are often pooled into a single virtual storage device, which is accessible by a consumer via the virtualization layer. Figure 1.2 depicts the layered structure of a simple storage model.

For storage systems, virtualization is a means by which multiple physical storage devices are viewed as a single logical unit. Virtualization can be accomplished in two ways at server level, fabric level, storage subsystem level, and file system level: *in-band* and *out-of-band* virtualization (Tate, 2003). In this

Figure 1.2 *Simple storage model (Bunn et al., 2004). HBA, Host Bus Adapter; RAID, Redundant Array of Independent Disks; LUN, Logical Unit Number.*

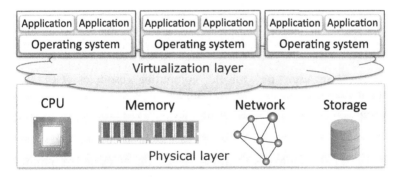

Figure 1.3 *Platform virtualization.*

context, in-band virtualization implies data and control flow over the same channel, whereas these flows are separated in out-of-band storage networks. This usually is achieved by separating data and metadata into different locations. An advantage of the latter approach is the availability of the full storage area network bandwidth to input/output (I/O) requests.

Platform Virtualization. The virtualization technologies discussed so far focus on resource virtualization. In contrast, platform virtualization embraces the complete computing infrastructure. Platform virtualization introduces a new layer between the physical hardware of the computer and the operating system layer (Fig. 1.3). This additional layer allows not only the coexistence but also the parallel execution of multiple operating system instances.

First and foremost, resource providers benefit from platform virtualization by reduced hardware, maintenance, and energy costs. Higher quality of service in terms of redundancy and security can also be achieved. For instance, platform virtualization allows the eviction of applications from physical nodes to prepare for maintenance. This can be achieved transparently and without interruption to the individual applications if they are running in virtual machines (Bhatia and Vetter, 2008).

End users of virtual machines benefit, on the one hand, from a flexible and fast deployment and, on the other hand, from a predictable and "clean" environment. Therefore, virtualization has become very popular for software testing and deployment.

1.4   CLOUD COMPUTING

Since 2007, cloud computing has become a new buzzword in the computer and information technology world. Vaquero et al. (2009) studied various existing definitions and tried, with limited success, to extract a set of attributes shared by all the definitions for cloud computing. There seems to be a reasonable consensus that the following characteristics are usually associated with cloud computing:

- **Scalability:** the ability to handle the growing use or the number of resources in a graceful manner or the ability to add resources in a seamless way
- **Pay per Use:** users pay for the computing resources (CPU, storage, network bandwidth, etc.) they consume
- **Virtualization:** providing an abstracted (or virtual) view of computing resources

Virtualization technologies help to generate the "illusion" of infinite computing and storage resources in a cloud computing environment. Depending on the services offered by a cloud, we can distinguish three types of clouds: Software as a service (SaaS), Platform as a service (PaaS), and Infrastructure as a service (IaaS). Figure 1.4 illustrates these in relation to the hardware layer.

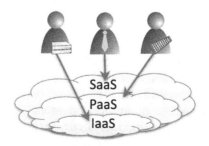

Figure 1.4   *Overview of "as-a-service" types.*

SaaS. This means that software is made available on demand to the end user by a distributed or decentralized cloud computing environment instead of a local computer. Depending on the available infrastructure and technology, one of two methods is applied to provide the software. The conventional method uses a terminal server and one or more clients. The software is installed at the server site. For access, users connect via the client to the server. The contemporary method to provide a SaaS is to deliver the service as part of a Web application; hence, access is normally through a Web browser.

Keeping the software at a central repository reduces the total overhead for maintenance (e.g., installation and updating); instead of several software installations, only one is maintained. The use of SaaS requires a reliable and fast link to the cloud provider. In some cases, a local installation is still preferable to improve performance.

One of the established SaaS providers is Google. Google Docs,[14] for example, offers an online word processing and spreadsheet software.

PaaS. PaaS environments (e.g., Google App Engine[15]) are used for the development and deployment of applications while avoiding the need to buy and manage the physical infrastructure. Compared to SaaS, PaaS provides facilities like database integration, application versioning, persistence, scalability, and security. Together, these services support the complete life cycle of developing and delivering Web applications and services available entirely through the Internet.

IaaS. IaaS defines the highest service level in privacy and reliability that resource providers currently offer their customers. In this case, the term infrastructure includes virtual and physical resources.

With respect to the interfaces offered by a cloud, a distinction is made between compute clouds and storage clouds. Compute cloud providers employ, for example, virtualization and by doing so offer easy access to remote computing power. In contrast to this, storage clouds focus on the persistent storage of data. Most compute clouds also offer interfaces to a storage facility which can be used to upload virtual appliances.

To satisfy the increasing demand for dynamic resource provisioning, the number of cloud providers is increasing steadily. Established providers are Amazon EC2,[16] FlexiScale,[17] and ElasticHosts.[18]

1.4.1 Drawbacks of Cloud Computing

At the peak of its hype, cloud computing was the proclaimed successor of grid computing. Unfortunately, cloud computing suffers from the same problems

---

[14] http://docs.google.com.
[15] http://code.google.com/intl/de/appengine/.
[16] http://aws.amazon.com/ec2/.
[17] www.flexiant.com/products/flexiscale/.
[18] www.elastichosts.com.

as grid computing—the difference is that in the cloud paradigm, the problems are located closer to the hardware layer.

The resource broker in a grid matches the job requirements (e.g., operating system and applications) with available resources and selects the most adequate resource for job[19] submission. In the context of cloud computing, the understanding of the term *job* must be revised to include virtual appliances. Nevertheless, a cloud customer is interested in deploying his job at the most adequate resource, so a matchmaking process is required. Without the existence of a cloud resource broker, the matchmaking is carried out manually.

A consequence caused by a missing cloud resource broker and by the natural human behavior to prefer favorite service providers often leads to a *vendor lock-in*.[20] The severity of the vendor lock-in problem is increased by proprietary and hence incompatible platform technologies at cloud provider level. For customers with data-intensive applications, it is difficult and expensive to migrate to a different cloud provider.[21] With the vendor lock-in, the customers strongly depend on the provided quality of service in a cloud (e.g., availability, reliability, and security). If a cloud suffers from poor availability, this implies poor availability of the customer's services.

A problem already present in grid computing is the limited network bandwidth between the data location and the computing resource. To bypass this bottleneck, commercial cloud providers started to physically mail portable storage devices to and from customers. For truly massive volumes of data, this "crude" mode of transferring data is relatively fast and cost-effective.[22]

1.4.2   Cloud Interfaces

A few years ago, standardization became one of the key factors in grid computing. Middleware like gLite and UNICORE have adopted open standards such as the Open Grid Services Architecture (OGSA) (Basic Execution Services) (Foster et al., 2008) and the Job Submission Description Language (JSDL) (Anjomshoaa et al., 2005). For cloud computing, it is not apparent if providers are interested in creating a standardized API. This API would ease the migration of data and services among cloud providers and results in a more competitive market.

Analyzing APIs of various cloud providers reveals that a common standard is unattainable in the foreseeable future. Many providers offer an API similar to the one of Amazon EC2 because of its high acceptance among customers.

At the moment, API development goes into two directions. The first direction is very similar to the developments for platform virtualization: The overlay

---

[19] A grid *job* is a binary executable or command to be submitted and run in a remote resource (machine) called server or "gatekeeper."

[20] A vendor lock-in makes a customer dependent on a vendor for products and services, unable to use another vendor without substantial switching costs.

[21] Data transfers within a cloud are usually free, but in- and outgoing traffic are charged.

[22] http://aws.amazon.com/importexport/.

API `libvirt`[23] was created, which supports various hypervisors (e.g., Xen, KVM, VMware, and VirtualBox). For cloud computing, there is the `libcloud` project[24]; the `libcloud` library hides inhomogeneous APIs of cloud providers (e.g., Slicehost, Rackspace, and Linode) from the user. The second direction tends toward a commonly accepted, standardized API for cloud providers. This API would make the development of `libcloud` needless.

The open cloud computing interface (OCCI) is one of the proposed API standards for cloud providers. It is targeted at providing a common interface for the management of resources hosted in IaaS clouds. OCCI allows resource aggregators to use a single common interface to multiple cloud resources and customers to interact with cloud resources in an ad hoc way (Edmonds et al., 2009).

A client encodes the required resources in a URL (see Fig. 1.5). Basic operations for resource modifications (create, retrieve, update, and delete) are mapped to the corresponding http methods `POST`, `GET`, `PUT`, and `DELETE` (Fielding et al., 1999). For example, a `POST` request for the creation of a computing resource looks similar to

```
POST /compute HTTP/1.1
Host: one.grid.tu-dortmund.de
Content-Length: 36
Content-Type: application/x-www-form-urlencoded
compute.cores=8&compute.memory=16384
```

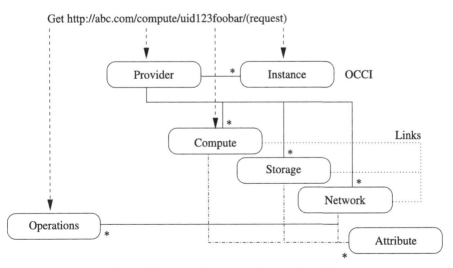

Figure 1.5 *Open cloud computing interface API.*

---

[23] http://libvirt.org/index.html.
[24] http://libcloud.apache.org.

Issuing this request triggers the creation of a virtual machine consisting of 8 cores and 16 GB of memory. Other attributes that can be requested are the CPU architecture (e.g., x86), a valid DNS name for the host, and the CPU speed in gigahertz.

The provision of storage (e.g., a virtual hard disk drive) via OCCI requires the specification of the storage size in gigabyte. Users are able to query the status (online, off-line, or "degraded") of the virtual storage, to back up, to resize, or for snapshot creation.

1.5   GRID AND CLOUD: TWO COMPLEMENTARY TECHNOLOGIES

Both grid and cloud computing are state-of-the-art technologies for large-scale computing. Grid and cloud computing could be viewed as mutually complementary technologies. Grid middleware is not likely to be replaced by cloud middleware because a cloud typically encompasses resources of only a single provider, while a grid spans resources of multiple providers. Nevertheless, compared with a grid resource, a cloud resource is more flexible because it adapts dynamically to the user requirements. Not surprisingly, one of the first "scientific" tasks clouds were used for was the deployment of virtual appliances containing grid middleware services. In the first days of cloud computing, only public, pay-per-use clouds (e.g., Amazon EC2) were available. Therefore, this endeavor was carried out only (1) to show its feasibility and (2) for benchmarking cloud capabilities of commercial providers. The added value of cloud computing was enough stimulus to develop open source computing and storage cloud solutions (e.g., OpenNebula,[25] Eucalyptus[26]).

With grid *and* cloud computing to their disposal, national and international e-science initiatives (e.g., D-Grid[27]) are currently reviewing their activities, which are mainly focused on grid computing. One aspect of D-Grid is to provide discipline-specific environments for collaboration and resource sharing to researchers. Independent from the discipline, a common core technology should be utilized. With the advent of cloud computing and its success in recent years, it is planned to be added to the already existing core technology, namely, grid computing.

In this context, two trends toward interoperation of grids and clouds are emerging. The first is implied by the lessons learned so far from grid computing and refers to the creation of a grid of clouds. Similar to single computing and storage resources gathered in a grid, computing and storage resources of a cloud will become part of a larger structure. As briefly shown in Section 1.4, additional value-added services like a cloud resource broker can be part of such a structure. In a possible grid of cloud setup, an information system peri-

---

[25] http://opennebula.org/.
[26] http://open.eucalyptus.com/.
[27] www.d-grid.de.

odically queries cloud resources, for example, health status, pricing information, and free capacities. This information is used by a cloud broker to find the most adequate resource to satisfy the user's needs. The specification of a common cloud API standard would ease the task of creating such a cloud broker.

The second identified trend started with a gap analysis. A set of components required for the seamless integration of cloud resources into existing grid infrastructures will be the result of this analysis. Using D-Grid as an example for a national grid initiative, work in the fields of user management and authentication, information systems, and accounting have been identified. Usually, authentication in a grid is based on a security infrastructure using X.509 certificates and/or short-lived credentials. Some commercial cloud providers offer a similar interface accepting certificates, but often, a username/password combination (e.g., `EC2_ACCESS_KEY` and `EC2_SECRET_KEY` for Amazon EC2) is employed for authentication.

Concerning the information systems, in D-Grid, each of the supported grid middlewares runs a separate one. The gLite middleware uses a combination of site and top-level BDII; (Web-)MDS is used in Globus Toolkit; and Common Information Service (CIS) is used in UNICORE6. The information provided by these systems is collected and aggregated by D-MON, a D-Grid monitoring service. D-MON uses an adapter for each type of supported grid middleware that converts the data received to an independent data format for further processing. To integrate information provided by a cloud middleware, a new adapter and the definition of measurement categories need to be developed.

After the creation of the missing components, grid and clouds are likely to coexist as part of e-science infrastructures.

1.6 MODELING AND SIMULATION OF GRID AND CLOUD COMPUTING

Today, modeling and simulation of large-scale computing systems are considered as a fruitful R&D area for algorithms and applications. With the increasing complexity of such environments, simulation technology has experienced great changes and has dealt with multiplatform distributed simulation, joining performance, and structure. A high-performance simulator is necessary to investigate various system configurations and to create different application scenarios before putting them into practice in real large-scale distributed computing systems. Consequently, simulation tools should enable researchers to study and evaluate developed methods and policies in a repeatable and controllable environment and to tune algorithm performance before deployment on operational systems. A grid or a cloud simulator has to be able to model heterogeneous computational components; to create resources, a network topology, and users; to enable the implementation of various job scheduling algorithms; to manage service pricing; and to output statistical results.

The existing discrete-event simulation studies cover a small number of grid sites, with only a few hundreds of hosts, as well as large-scale use case grids, including thousands or millions of hosts. Common simulation studies deal with job scheduling and data replication algorithms to achieve better performance and high availability of data.

A list of grid simulation tools that implement one or more of the above-mentioned functionalities include the following:

1. *OptorSim* (Bell et al., 2002) is being developed as part of the EU DataGrid project. It mainly emphasizes grid replication strategies and optimization.

2. *SimGrid* toolkit (Casanova, 2001), developed at the University of California at San Diego, is a C-based simulator for scheduling algorithms.

3. *MicroGrid* emulator (Song et al., 2000), developed at the University of California at San Diego, can be used for building grid infrastructures. It allows to execute applications, created using the Globus Toolkit, in a virtual grid environment.

4. *GangSim* (Dumitrescu and Foster, 2005), developed at the University of Chicago, aims to study the usage and scheduling policies in a multivirtual organization environment and is able to model real grids of considerable size.

1.6.1 GridSim and CloudSim Toolkits

GridSim (Sulistio et al., 2008) is a Java-based discrete-event grid simulation toolkit, developed at the University of Melbourne. It enables modeling and simulation of heterogeneous grid resources, users, and applications. Users can customize algorithms and workload.

At the lowest layer of the GridSim architecture operates the SimJava (Howell and McNab, 1998) discrete-event simulation engine. It provides GridSim with the core functionalities needed for higher-level simulation scenarios, such as queuing and processing of events, creation of system components, communication between components, and management of the simulation clock.

The GridSim toolkit enables to run reproducible scenarios that are not possible in a real grid infrastructure. It supports high-level software components for modeling multiple grid infrastructures and basic grid components such as the resources, workload traces, and information services. GridSim enables the modeling of different resource characteristics. Therefore, it allocates incoming jobs based on space or time-shared mode. It is able to schedule compute- and data-intensive jobs; it allows easy implementation of different resource allocation algorithms; it supports reservation-based or auction mechanisms for resource allocation; it enables simulation of virtual organization

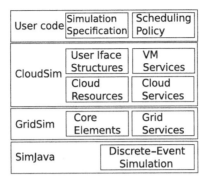

Figure 1.6 *CloudSim architecture layers. Iface, interface; VM, virtual machine.*

scenarios; it is able to visualize tracing sequences of simulation execution; and it bases background network traffic characteristics on a probabilistic distribution (Buyya and Murshed, 2002).

CloudSim (Buyya et al., 2009) is a framework, also developed at the University of Melbourne, that enables modeling, simulation, and experimenting on cloud computing infrastructures. It is built on top of GridSim (Fig. 1.6).

CloudSim is a platform that can be used to model data centers, service brokers, scheduling algorithms, and allocation policies of a large-scale cloud computing platform. It supports the creation of virtual machines on a simulated node, cloudlets, and allows assigning jobs to appropriate virtual machines. It also enables simulation of various data centers to allow investigations on federation and related policies for the migration of virtual machines.

1.7   SUMMARY AND OUTLOOK

The demand for computing and storage resources has been increasing steadily over many years. In the past, static resources of multiple providers were subsumed into a grid. The concept of resource sharing within a grid and its drawbacks were described in Section 1.2. Section 1.3 introduced two different virtualization types: resource and platform virtualization. Especially, platform virtualization is one of the key enabling technologies for the concept of cloud computing. Cloud providers act as service providers and offer to customers software (SaaS), development platforms (PaaS), and infrastructure (IaaS). Section 1.4 provided a brief overview of cloud computing and the three main service types (SaaS, PaaS, and IaaS).

Section 1.5 provided an outlook on efforts attempting to integrate cloud and grid middleware. e-Science initiatives need to provide a uniform and simple interface to both types of resources to facilitate future developments.

Section 1.6 discussed the need to model and simulate grid and cloud environments and important grid simulation toolkits (e.g., the Java-based simulators GridSim and CloudSim).

REFERENCES

L. Abadie, P. Badino, J. P. Baud, et al. Grid-enabled standards-based data management. In *24th IEEE Conference on Mass Storage Systems and Technologies*, pp. 60–71, Los Alamitos, California, IEEE Computer Society. 2007.

C. Aiftimiei, M. Sgaravatto, L. Zangrando, et al. Design and implementation of the gLite CREAM job management service. *Future Generation Computer Systems*, 26(4):654–667, 2010.

A. Anjomshoaa, F. Brisard, M. Drescher, et al. Job Submission Description Language (JSDL) Specification: Version 1.0, 2005. http://www.gridforum.org/documents/GFD.56.pdf.

P. Badino, O. Barring, J. P. Baud, et al. The storage resource manager interface specification version 2.2, 2009. http://sdm.lbl.gov/srm-wg/doc/SRM.v2.2.html.

W. Bell, D. Cameron, L. Capozza, et al. Simulation of dynamic grid replication strategies in OptorSim. In M. Parashar, editor, *Grid Computing, Volume 2536 of Lecture Notes in Computer Science*, pp. 46–57, Berlin: Springer, 2002.

N. Bhatia and J. Vetter. Virtual cluster management with Xen. In L. Bougé, M. Alexander, S. Childs, et al., editors, *Euro-Par 2007 Workshops: Parallel Processing, volume 4854 of Lecture Notes in Computer Science*, pp. 185–194, Berlin and Heidelberg: Springer-Verlag, 2008.

F. Bunn, N. Simpson, R. Peglar, et al. Storage virtualization: SNIA technical tutorial, 2004. http://www.snia.org/sites/default/files/sniavirt.pdf.

S. Burke, S. Campana, E. Lanciotti, et al. gLite 3.1 user guide: Version 1.2, 2009. https://edms.cern.ch/file/722398/1.2/gLite-3-UserGuide.html.

R. Buyya and M. M. Murshed. GridSim: A toolkit for the modeling and simulation of distributed resource management and scheduling for grid computing. *Concurrency and Computation: Practice and Experience*, 14(13–15):1175–1220, 2002.

R. Buyya, R. Ranjan, and R. N. Calheiros. Modeling and simulation of scalable cloud computing environments and the CloudSim Toolkit: Challenges and opportunities. In Waleed W. Smari and John P. McIntire, editor, *Proceedings of the 2009 International Conference on High Performance Computing and Simulation*, pp. 1–11, Piscataway, NJ: IEEE, 2009.

H. Casanova. Simgrid: A toolkit for the simulation of application scheduling. In R. Buyya, editor, *CCGRID'01: Proc. of the 1st Int'l Symposium on Cluster Computing and the Grid*, pp. 430–441, Los Alamitos, CA: IEEE Computer Society, 2001.

C. L. Dumitrescu and I. Foster. GangSim: A simulator for grid scheduling studies. In *CCGRID'05: Proc. of the 5th IEEE Int'l Symposium on Cluster Computing and the Grid*, pp. 1151–1158, Washington, DC: IEEE Computer Society, 2005.

A. Edmonds, S. Johnston, G. Mazzaferro, et al. Open Cloud Computing Interface Specification version 5, 2009. http://forge.ogf.org/sf/go/doc15731.

R. Fielding, J. Gettys, J. Mogul, et al. RFC 2616: Hypertext transfer protocol—HTTP/1.1. *Status: Standards Track*, 1999. http://www.ietf.org/rfc/rfc2616.txt.

I. Foster. What is the grid?—A three point checklist. *GRIDToday*, 1(6):22–25, 2002.

I. Foster. Globus Toolkit version 4: Software for service-oriented systems. In *Network and parallel computing*, pp. 2–13, 2005.

I. Foster, A. Grimshaw, P. Lane, et al. OGSA basic execution service: Version 1.0, 2008. http://www.ogf.org/documents/GFD.108.pdf.

I. Foster, C. Kesselman, and S. Tuecke. The anatomy of the grid: Enabling scalable virtual organizations. *Int'l Journal of High Performance Computing Applications*, 15(3):200–222, 2001.

F. Howell and R. McNab. SimJava: A discrete event simulation library for Java. In *Int'l Conference on Web-Based Modeling and Simulation*, pp. 51–56, 1998.

IEEE Computer Society. IEEE standard for local and metropolitan area networks virtual bridged local area networks, 2006. http://ieeexplore.ieee.org.

A Mowshowitz. Virtual organization. *Communications of the ACM*, 40(9):30–37, 1997.

H. J. Song, X. Liu, D. Jakobsen, et al. The MicroGrid: A scientific tool for modeling computational grids. *Scientific Programming*, 8(3):127–141, 2000.

A. Sulistio, R. Buyya, U. Cibej, et al. A toolkit for modelling and simulating data grids: An extension to GridSim. *Concurrency and Computation: Practice and Experience*, 20(13):1591–1609, 2008.

J. Tate. Virtualization in a SAN, 2003. http://www.redbooks.ibm.com/redpapers/pdfs/redp3633.pdf.

J. E. Thornton. Parallel operation in the control data 6600. In *AFIPS 1964 Fall Joint Computer Conference*, pp. 33–41, Spartan Books, 1965.

L. M. Vaquero, L. Rodero-Merino, J. Caceres, et al. A break in the clouds: Towards a cloud definition. *ACM SIGCOMM Computer Communication Review*, 39(1):50–55, 2009.

Chapter 2

The e-Infrastructure Ecosystem: Providing Local Support to Global Science

Erwin Laure and Åke Edlund

KTH Royal Institute of Technology, Stockholm, Sweden

2.1 THE WORLDWIDE E-INFRASTRUCTURE LANDSCAPE

Modern science is increasingly dependent on information and communication technologies (ICTs), analyzing huge amounts of data (in the terabyte and petabyte range), running large-scale simulations requiring thousands of CPUs (in the teraflop and petaflop range), and sharing results between different research groups. This collaborative way of doing science has led to the creation of virtual organizations that combine researches and resources (instruments, computing, and data) across traditional administrative and organizational domains (Foster et al., 2001). Advances in networking and distributed computing techniques have enabled the establishment of such virtual organizations, and more and more scientific disciplines are using this concept, which is also referred to as *grid computing* (Foster and Kesselman, 2003; Lamanna and Laure, 2008). The past years have shown the benefit of basing grid computing on a well-managed

Large-Scale Computing, First Edition. Edited by Werner Dubitzky, Krzysztof Kurowski, Bernhard Schott.
© 2012 John Wiley & Sons, Inc. Published 2012 by John Wiley & Sons, Inc.

infrastructure federating the network, storage, and computing resources across a big number of institutions and making them available to different scientific communities via well-defined protocols and interfaces exposed by a software layer (grid middleware). This kind of federated infrastructure is referred to as e-infrastructure. Europe is playing a leading role in building multinational, multidisciplinary e-infrastructures. Initially, these efforts were driven by academic proof-of-concept and test-bed projects, such as the European Data Grid Project (Gagliardi et al., 2006), but have since developed into large-scale, production e-infrastructures supporting numerous scientific communities. Leading these efforts is a small number of large-scale flagship projects, mostly cofunded by the European Commission, which take the collected results of predecessor projects forward into new areas. Among these flagship projects, the Enabling Grids for E-sciencE (EGEE) project unites thematic, national, and regional grid initiatives in order to provide an e-infrastructure available to all scientific research in Europe and beyond in support of the European Research Area (Laure and Jones, 2009). But EGEE is only one of many e-infrastructures that have been established all over the world. These include the U.S.-based Open Science Grid (OSG)[1] and TeraGrid[2] projects, the Japanese National Research Grid Initiative (NAREGI)[3] project, and the Distributed European Infrastructure for Supercomputing Applications (DEISA)[4] project, a number of projects that extend the EGEE infrastructure to new regions (see Table 2.1), as well as (multi)national efforts like the U.K. National Science Grid (NGS),[5] the Northern Data Grid Facility (NDGF)[6] in northern Europe,

TABLE 2.1   Regional e-Infrastructure Projects Connected to EGEE

Project	Web Site	Countries Involved
BalticGrid	http://www.balticgrid.eu	Estonia, Latvia, Lithuania, Belarus, Poland, Switzerland, and Sweden
EELA	http://www.eu-eela.org	Chile, Cuba, Italy
		Argentina, Brazil
		Mexico, Peru, Portugal, Spain, Venezuela
EUChinaGrid	http://www.euchinagrid.eu	China, Greece, Italy, Poland, and Taiwan
EUIndiaGrid	http://www.eumedgrid.eu	India, Italy, and United Kingdom
EUMedGrid	http://www.euindiagrid.eu	Algeria, Cyprus, Egypt, Greece, Jordan
		Israel, Italy, Malta, Morocco, Palestine
		Spain, Syria, Tunisia, Turkey, United Kingdom
SEE-GRID	http://www.see-grid.eu	Croatia, Greece, Hungary, FYR of Macedonia
		Albania, Bulgaria, Bosnia and Herzegovina
		Moldova, Montenegro, Romania, Serbia, Turkey

[1] http://www.opensciencegrid.org.
[2] http://www.teragrid.org.
[3] http://www.naregi.org.
[4] http://www.deisa.org.
[5] http://www.ngs.ac.uk.
[6] http://www.ndgf.org.

and the German D-Grid.[7] Together, these projects cover large parts of the world and a wide variety of hardware systems.

In all these efforts, providing support and bringing the technology as close as possible to the scientist are of utmost importance. Experience has shown that users require local support when dealing with new technologies rather than a relatively anonymous European-scale infrastructure. This local support has been implemented, for instance, by EGEE through regional operation centers (ROCs) as well as DEISA, which assigns *home* sites to users, which are providing user support and a base storage infrastructure. In a federated model, local support can be provided by national or regional e-infrastructures, as successfully demonstrated by the national and regional projects mentioned earlier. These projects federate with international projects like EGEE to form a rich infrastructure ecosystem under the motto "think globally, act locally." In the remainder of this chapter, we exemplify this strategy on the BalticGrid and EGEE projects, discuss the impact of clouds on the ecosystem, and provide an outlook on future developments.

2.2 BALTICGRID: A REGIONAL E-INFRASTRUCTURE, LEVERAGING ON THE GLOBAL "MOTHERSHIP" EGEE

To establish a production-quality e-infrastructure in a greenfield region like the Baltic region, a dedicated project was established, the BalticGrid[8] project. According to the principle of think globally, act locally, the aim of the project was to build a regional e-infrastructure that seamlessly integrates with the international e-infrastructure of EGEE. In the first phase of the BalticGrid project, starting in 2005, the necessary network and middleware infrastructure was rolled out and connected to EGEE. The main objective of BalticGrid's second phase, BalticGrid-II (2008–2010), was to further increase the impact, adoption, and reach of e-science infrastructures to scientists in the Baltic States and Belarus, as well as to further improve the support of services and users of grid infrastructures. As with its predecessor BalticGrid, BalticGrid-II continued with strong links with EGEE and its technologies, with gLite (Laure et al., 2006) as the underlying middleware of the project's infrastructure.

2.2.1 The BalticGrid Infrastructure

The BalticGrid infrastructure is in production since 2006 and has been used significantly by the regional scientific community. The infrastructure consists of 26 clusters in five countries, of which 18 are on the EGEE production infrastructure, with more than 3500 CPU cores, 230 terabytes of storage space.

One of the first challenges of the BalticGrid project was to establish a reliable network in Estonia, Latvia, and Lithuania as well as to ensure optimal

---

[7] http://www.d-grid.org.
[8] http://www.balticgrid.eu.

network performance for large file transfers and interactive traffic associated with grids. This was successfully achieved in the projects in the first year exploiting the European GéANT network infrastructure, adding Belarus in the second phase.

2.2.2 BalticGrid Applications: Providing Local Support to Global Science

The resulting BalticGrid infrastructure supports and helps scientists from the region to foster the use of modern computation and data storage systems, enabling them to gain knowledge and experience to work in the European research area.

The main application areas within the BalticGrid are from high-energy physics, materials science and quantum chemistry, framework for engineering modeling tasks, bioinformatics and biomedical imaging, experimental astrophysical thermonuclear fusion (in the framework of the ITER project), linguistics, as well as operational modeling of the Baltic Sea ecosystem.

To support these, a number of applications have been ported, often leveraging on earlier EGEE work, to the BalticGrid:

- **ATOM:** a set of computer programs for theoretical studies of atoms
- **Complex Comparison of Protein Structures:** an application that offers a method applied for the exploration of potential evolutionary relationships between the CATH protein domains and their characteristics
- **Crystal06:** a quantum chemistry package to model periodic systems
- **Computational Fluid Dynamics (FEMTOOL):** modeling of viscous incompressible free surface flows
- **DALTON 2.0:** a powerful molecular electronic structure program with extensive functions for the calculation of molecular properties at different levels of theory
- **Density of Montreal:** a molecular electronic structure program
- **ElectroCap Stellar Rates of Electron Capture:** a set of computer codes produce nuclear physics input for core-collapse supernova simulations
- **Foundation Engineering (Grill):** global optimization of grillage-type foundations using genetic algorithms
- **MATLAB:** distributed computing server
- **MOSES SMT Toolkit (with SRILM):** a factored phrase-based beam search decoder for machine translators
- **NWChem:** a computational chemistry package
- **Particle Technology (DEMMAT):** particle flows, nanopowders, and material structure modeling on a microscopic level using the discrete element method

- **Polarization and Angular Distributions in Electron-Impact Excitation of Atoms (PADEEA)**
- **Vilnius Parallel Shell Model Code:** an implementation of nuclear spherical shell model approach

2.2.3 The Pilot Applications

To test, validate, and update the support model, a smaller set of pilot applications was chosen. These applications received special attention during the initial phase of BalticGrid-II and helped shape BalticGrid's robust and cost-efficient overall support model.

2.2.3.1 Particle Technology (DEMMAT) The development of the appropriate theoretical framework as well as numerical research tools for the prediction of constitutive behavior with respect to microstructure belongs to major problems of computational sciences. In general, the macroscopic material behavior is predefined by the structure of grains of various sizes and shapes, or even by individual molecules or atoms. Their motion and interaction have to be taken into account to achieve a high degree of accuracy. The discrete element method is an attractive candidate to be applied for modeling granular flows, nanopowders, and other materials on a microscopic level. It belongs to the family of numerical methods and refers to conceptual framework on the basis of which appropriate models, algorithms, and computational technologies are derived (Fig. 2.1).

The main disadvantages of the discrete element method technique, in comparison to the well-known continuum methods, are related to computational capabilities that are needed to handle a large number of particles and a short time interval of simulations. The small time step imposed in the explicit time integration schemes gives rise to the requirement that a very large number of time increments should be performed. Grid and distributed computing technologies are a standard way to address industrial-scale computing problems.

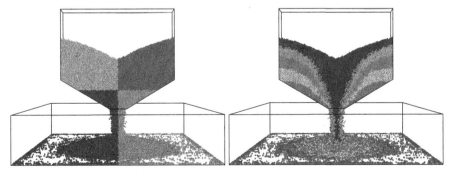

Figure 2.1 *Particle flow during hopper discharge. Particles are colored according to resultant force.*

Interdisciplinary cooperation and development of new technologies are major factors driving the progress of discrete element method models and their countless applications (e.g., nanopowders, compacting, mixing, hopper discharge, and crack propagation in building constructions).

2.2.3.2 Materials Science, Quantum Chemistry

NWChem[9] is a computational chemistry package that has been developed by the Molecular Sciences Software group of the Environmental Molecular Sciences Laboratory at the Pacific Northwest National Laboratory, USA.

NWChem provides many methods to compute the properties of molecular and periodic systems using standard quantum mechanical descriptions of the electronic wave function or density. In addition, NWChem has the capability to perform classical molecular dynamics and free energy simulations. These approaches may be combined to perform mixed quantum mechanics and molecular mechanics simulations.

2.2.3.3 CORPLT: Corpus of Academic Lithuanian

The corpus (large and structured set of texts) was designed as a specialized corpus for the study of academic Lithuanian, and it will be the first synchronic corpus of academic written Lithuanian in Lithuania. It will be a major resource of authentic language data for linguistic research of academic discourse, for interdisciplinary studies, lexicographical practice, and terminology studies in theory and practice. The compilation of the corpus will follow the most important criteria: methods, balance, representativeness, sampling, TEI P5 Guidelines, and so on. The grid application will be used for testing algorithms of automatic encoding, annotation, and search-analysis steps. Encoding covers recognition of text parts (sections, titles, etc.) and correcting of text flow. Linguistic annotation consists of part of speech tagging, part of sentence tagging, and so on. The search-analysis part deals with the complexity level of the search and tries to distribute and effectively deal with the load for corpus services.

2.2.3.4 Complex Comparison of Protein Structures

The Complex Comparison of Protein Structures application supports the exploration of potential evolutionary relationships between the CATCH protein domains and their characteristics and uses an approach called *3D graphs*. This tool facilitates the detection of structural similarities as well as possible fold mutations between proteins.

The method employed by the Complex Comparison of Protein Structures tool consists of two stages:

- **Stage 1:** all-against-all comparison of CATCH domains by the ESSM software
- **Stage 2:** construction of fold space graphs on the basis of the output of the first stage

[9] http://www.nwchem-sw.org.

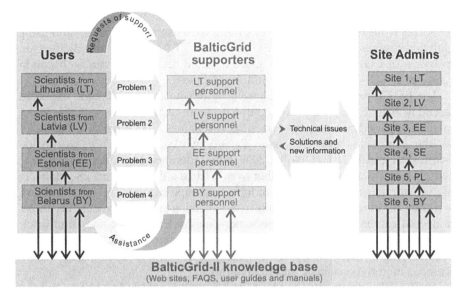

Figure 2.2   *The BalticGrid support model.*

This method is used for evolutionary aspects of protein structure and function. It is based on the assumption that protein structures, similar to sequences, have evolved by a stepwise process, each step involving a small change in protein fold. The application ("gridified" within BalticGrid-II) accesses the Protein DataBase to compare proteins individually and in parallel.

2.2.4   BalticGrid's Support Model

By working with these pilot applications, a robust, scalable, and cost-efficient support model could be developed. Although usually seen as a homogeneous region, the Baltic states expose significant differences, with different languages, different cultures, and different ways local scientists use to interact with their resource providers. Due to these differences, local support structures in the individual countries have been established to provide users with the best possible local support. These local support teams interact with resource providers irrespective of their origin and build upon a joint knowledge base consisting of Web sites, FAQs, user guides, and manuals that is constantly enriched by the support teams. This knowledge base is also available to the users, reducing the need to contact support. The overall support model is depicted in Figure 2.2.

2.3   THE EGEE INFRASTRUCTURE

The BalticGrid example mentioned earlier is to be seen as an extension of EGEE, given that a large number of BalticGrid sites are EGEE certified and are utilized by the EGEE users. As a result, we have both global users using resources from BalticGrid, as well as BalticGrid users participating and using global resources from the full EGEE setting. To complete the picture, an overview of EGEE is given next; more detailed descriptions can be found in Laure et al. (2006) and in Laure and Jones (2009).

The EGEE project is a multiphase program starting in 2004 and expected to end in 2010. EGEE has built a pan-European e-infrastructure that is being increasingly used by a variety of scientific disciplines. EGEE has also expanded to the Americas and the Asia Pacific, working toward a worldwide e-infrastructure. EGEE (October 2009 data) federates more then 260 resource centers from 55 countries, providing over 150,000 CPU cores and 28 petabytes of disk storage and 41 petabytes of long-term tape storage. The infrastructure is routinely being used by over 5000 users forming some 200 virtual organizations and running over 330,000 jobs per day. EGEE users come from disciplines as diverse as archaeology, astronomy, astrophysics, computational chemistry, earth science, finance, fusion, geophysics, high-energy physics, life sciences, materials sciences, and many more.

The EGEE infrastructure consists of a set of services and test beds, and the management processes and organizations that support them.

2.3.1 The EGEE Production Service

The computing and storage resources EGEE integrates are provided by a large and growing number of resource centers, mostly in Europe but also in the Americas and the Asia Pacific.

EGEE has decided that a central coordination of these resource centers would not be viable mainly because of scaling issues. Hence, EGEE has developed a distributed operations model where ROCs take over the responsibility of managing the resource centers in their region. Regions are defined geographically and include up to eight countries. This setup allows adjusting the operational procedures to local peculiarities like legal constraints, languages, and best practices. Currently, EGEE has 11 ROCs managing between 12 and 38 sites. CERN acts as a special *catchall* ROC taking over responsibilities for sites in regions not covered by an existing ROC. Once the number of sites in that region has reached a significant number (typically over 10), the creation of a new ROC is envisaged. EGEE's ROC model has proven very successful, allowing the fourfold growth of the infrastructure (in terms of sites) to happen without any impact on the service.

ROCs are coordinated by the overall Operations Coordination located at CERN. Coordination mainly happens through weekly operations meetings where issues from the different ROCs are being discussed and resolved. A so-called grid operator on duty is constantly monitoring the health of the EGEE infrastructure using a variety of monitoring tools and initiates actions in case services or sites are not in a good state. EGEE requires personnel at resource centers that manage the grid services offered and take corrective measures in case of problems. Service-level agreements are being set up between EGEE and the resource centers to fully define the required level of commitment, which may differ between centers. The Operations Coordination is also responsible for releasing the middleware distribution for deployment; individual sites are supported by their ROC in installing and configuring the middleware.

Security is a cornerstone of EGEE's operation and the Operational Security Coordination Group is coordinating the security-related operations at all EGEE sites. In particular, the EGEE Security Officer interacts with the site's security contacts and ensures security-related problems are properly handled, security policies are adhered to, and the general awareness of security-related issues is raised. EGEE's security policies are defined by the Joint Security Policy Group, a group jointly operated by EGEE, the U.S. OSG project, and the Worldwide LHC Computing Grid project (LCG).[10] Further security-related policies are set by the EUGridPMA and the International Grid Trust Federation (IGTF),[11] the bodies approving certificate authorities and thus establishing trust between different actors on the EGEE infrastructure. A dedicated Grid Security Vulnerability Group proactively analyzes the EGEE infrastructure and its services to detect potential security problems early on and to initiate their remedy. The EGEE security infrastructure is based on X.509 proxy certificates which allow to implement the security policies both at site level as well as on grid-wide services.

The EGEE infrastructure is federating resources and making them easily accessible but does not own the resources itself. Instead, the resources accessible belong to independent resource centers that procure their resources and allow access to them based on their particular funding schemes and policies. Federating the resources through EGEE allows the resource centers to offer seamless, homogeneous access mechanisms to their users as well as to support a variety of application domains through the EGEE virtual organizations. Hence, EGEE on its own cannot take any decision on how to assign resources to virtual organizations and applications. EGEE merely provides a marketplace where resource providers and potential users negotiate the terms of usage on their own. EGEE facilitates this negotiation in terms of the Resource Access Group, which brings together application representatives and the ROCs. Overall, about 20% of the EGEE resources are provided by nonpartner organizations, and while the majority of the EGEE resource centers are academic institutions, several industrial resource centers participate on the EGEE infrastructure to gain operational experience and to offer their resources to selected research groups.

All EGEE resources centers and particularly ROCs are actively supporting their users. As it is sometimes difficult for a user to understand which part of the EGEE infrastructure is responsible for a particular support action, EGEE operates the Global Grid User Support (GGUS). This support system is used

---

[10] http://cern.ch/lcg.
[11] http://www.gridpma.org/.

throughout the project as a central entry point for managing problem reports and tickets, for operations, as well as for user, virtual organization, and application support. The system is interfaced to a variety of other ticketing systems in use in the regions/ROCs. This is in order that tickets reported locally can be passed to GGUS or other areas, and that operational problem tickets can be pushed down into local support infrastructures. Overall, GGUS deals on average with over 1000 tickets per month, half of them due to the ticket exchange with network operators as explained next.

All usage of the EGEE infrastructure is accounted and all sites collect usage records based on the Open Grid Forum (OGF)[12] usage record recommendations. These records are both stored at the site and pushed into a global database that allows retrieving statistics on the usage per virtual organization, regions, countries, and sites. Usage records are anonymized when pushed into the global database, which allows retrieving them via a Web interface.[13] The accounting data are also gradually used to monitor service-level agreements that virtual organizations have set up with sites.

EGEE also provides an extensive training program to enable users to efficiently exploit the infrastructure. A description of this training program is out of the scope of this chapter and the interested reader is referred to the EGEE training Web site[14] for further information.

An e-infrastructure is highly dependent on the underlying network provisioning. EGEE is relying on the European Research Network operated by DANTE and the National Research and Educational Networks. The EGEE Network Operations Center links EGEE to the network operations and ensures both EGEE and the network operations are aware of requirements, problems, and new developments. Particularly, the network operators push their trouble tickets into GGUS, thus integrating them into the standard EGEE support structures.

2.3.2 EGEE and BalticGrid: e-Infrastructures in Symbiosis

To fully leverage on the EGEE project success, BalticGrid has reused much of the work done with respect to operations, support organization, education, dissemination, and so forth, as well as all the work put into the EGEE middleware gLite. The goal in making as many BalticGrid sites EGEE certified, that is, seamless parts of the overall EGEE infrastructure, is to enable users from the Baltic and Belarus region to participate as much as possible in the global research projects.

From the EGEE perspective, BalticGrid is expanding the reach and increasing the total capacity of the infrastructure. In addition, BalticGrid has been a

---

[12] http://www.ogf.org.

[13] http://www3.egee.cesga.es/gridsite/accounting/CESGA/egee_view.php.

[14] http://www.egee.nesc.ac.uk.

test and meeting point for coexisting middlewares, gLite with ARC[15] and UNICORE.[16]

The larger project has helped the smaller to a very quick and economic launch. This was achieved by serving the smaller project with already prepared structures, policies, and middleware. The smaller project, being more flexible due to its size, has acted as a test bed for aspects such as interoperability of middlewares and the deployment and use of cloud computing for small and medium enterprises (SMEs) and start-up companies.

The cooperation of the two projects is intended to improve the environment for their users.

2.4   INDUSTRY AND E-INFRASTRUCTURES: THE BALTIC EXAMPLE

2.4.1   Industry and Grids

Industry and academia share many of the challenges in using distributed environments in cost-efficient and secure ways. To learn more from each other, there have been a number of initiatives to further engage the industry to participate in academic grid projects.

Examples are mainly from life science, materials science, but also economics, where financial institutions are using parts of the open source grid middleware developed by academic projects. In addition, there have been some spin-offs from academia resulting in enterprise software offerings, for example, the UNIVA[17] start-up company with founders from the academic and open source community.

The overall uptake of grid technology by small- and medium-sized companies has not been as high as first anticipated. One reason is possibly the complexity of the underlying problem—to share resources in a seamless fashion, introducing a number of complex techniques to ensure this functionality in a secure and controlled fashion. In many ways, the grid offering is very useful from a business perspective, for example, establishing new markets for resources, but to this day, still much work is needed from the user to join such environments, and much work is needed to port users' applications to use the grid systems. This is, to a large extent due to the academic focus of most grid projects that put higher weights on functionality and performance than on user experiences. The biggest obstacles encountered were the usage of X.509 certificates as authentication means as well as the different interfaces and semantics of grid services originating from different research groups. Standardization efforts (Riedel et al., 2009) aim at overcoming the latter problem. In addition, to share resources as a concept is still considered a big barrier in many industries also within corporations, and security, especially mutual trust issues, is

---

[15] http://www.nordugrid.org/middleware.
[16] http://www.unicore.eu.
[17] http://www.univaud.com.

usually the first obstacle mentioned. Another issue is the overall quality of the open source middlewares, still considered too low for many companies.

2.4.2 Industry and Clouds, Clouds and e-Infrastructures

In the commercial sector, dynamic resource and service provisioning as well as "pay-per-use" concepts are being pushed to the next level with the introduction of "cloud computing," successfully pioneered by Amazon with their Elastic Compute Cloud and Simple Storage Service offerings. Many other major IT businesses offer cloud services today, including Google, IBM, and Microsoft. Using virtualization techniques, these infrastructures allow dynamic service provisioning and give the user the illusion of having access to virtually unlimited resources on demand. This computing model is particularly interesting for start-up ventures with limited IT resources as well as for dynamic provisioning of additional resources to cope with peak demands, rather than overprovisioning one's own infrastructure. As of today, the usability of these commercial offerings for research remains yet to be shown, although a few promising experiments have already been performed. In principle, clouds could be considered yet another resource provider in e-infrastructures, with a need to bridge the different interfaces and operational procedures to provide the researcher with a seamless infrastructure. More work on interface standardization and on how commercial offerings can be made part of the operations of academic research infrastructures is needed.

2.4.3 Clouds: A New Way to Attract SMEs and Start-Ups

Compared to grid, in cloud services, the ambition and the resulting complexity is lower, resulting in a lower barrier for the user as well as higher security, availability, and quality. All these simplifications, with the high degree of flexibility, has caught great interest from the industry, and while open grid efforts are led by academia, cloud computing is, to a high degree, business driven.

For smaller companies, flexibility is the key to be able to quickly launch and delaunch, and to be able to move capital investments (buying hardware) to operation expenses (*hiring* infrastructure, pay-per-use). Larger companies are more hesitant, trying clouds in a stepwise manner, combining private clouds with public-clouds-when-needed, so called hybrid clouds.

First out of the smaller companies are the newly started ones, the start-ups. For many of these companies there are no alternatives but to use all means of cost minimization, and here cloud computing fits very well. Start-ups do not have much time in building their own infrastructure and usually do not really know what they will need in the near future, and here cloud computing offers a quick launch, as well as delaunch, when needed. Investors are also benefiting from this shift in cost models, lowering their initial risks and getting quick proof of concept, still with scalable solutions in case of early success.

2.4.3.1 BalticGrid Innovation Lab (BGi) Within the BalticGrid project, a focused effort to attract SMEs and start-ups to use the regional e-infrastructure was made, resulting in the creation of the BGi.[18] BGi aims to educate early-stage start-ups in the use of BalticGrid resources, mainly through a cloud interface—the BalticCloud.[19] BalticCloud is a cloud infrastructure based on open source solutions (Eucalyptus, OpenNebula) (Sotomayor et al., 2008; Nurmi et al., 2009). At BGi, companies learn about how to leverage on cloud computing, both to change their cost model and possibly to lower their own internal IT costs (Assuncao et al., 2009). Companies learn also how to prepare short-term pilots, prototyping, and novel services to their customers. BGi is also a business networking effort, all with BalticGrid in common.

2.4.3.2 Clouds and SMEs: Lessons Learned BGi, together with BalticCloud, shows a way to attract SMEs and start-ups in the region, and the experience so far is very positive. Early examples are from the movie production industry (rendering on BalticCloud) and larger IT infrastructure companies preparing cloud services for start-ups. The project has also produced a number of cloud consultancy firms, and there is a growing interest of cloud support for innovation services, for example, incubators, in the region.

2.5 THE FUTURE OF EUROPEAN E-INFRASTRUCTURES: THE EUROPEAN GRID INITIATIVE (EGI) AND THE PARTNERSHIP FOR ADVANCED COMPUTING IN EUROPE (PRACE) INFRASTRUCTURES

The establishment of a European e-infrastructure ecosystem is currently progressing along two distinct paths: EGI and PRACE. EGI intends to federate national and regional e-infrastructures, managed locally by National Grid Initiatives (NGIs) into a pan-European, general-purpose e-infrastructure as pioneered by the EGEE project that unites thematic, national, and regional grid initiatives. EGI is a direct result of the European e-Infrastructure Reflection Group (e-IRG) recommendation to develop a sustainable base for European e-infrastructures. Most importantly, funding schemes are being changed from short-term project funding (like 2 years' funding periods in the case of EGEE) to sustained funding on a national basis. This provides researchers with the long-term perspective needed for multiyear research engagements. All European countries support the EGI vision, with final launch during 2010. Unlike EGEE, which has strong central control, EGI will consist of largely autonomous NGIs with a lightweight coordination entity on the European level. This setup asks for an increased usage of standardized services and operational procedures to enable a smooth integration of different NGIs exposing a common layer to the user while preserving their own autonomy. In

---

[18] http://www.balticgrid.eu/bgi.
[19] http://cloud.balticgrid.eu.

this context, the work of the OGF is of particular importance. The goal of OGF is to harmonize these different interfaces and protocols and to develop standards that will allow interoperable, seamlessly accessible services. One notable effort in this organization where all of the above-mentioned projects work together is the Grid Interoperation Now (GIN) (Riedel et al., 2009) group. Through this framework, the mentioned infrastructures work to make their systems interoperate with one another through best practices and common rules. GIN does not intend to define standards but to provide the groundwork for future standardization performed in other groups of OGF. Already this has led to seamless interoperation between OSG and EGEE, allowing jobs to freely migrate between the infrastructures as well as seamless data access using a single sign-on. Similar efforts are currently ongoing with NAREGI, DEISA, and NDGF. As part of the OGF GIN effort, a common service discovery index of nine major grid infrastructures worldwide has been created, allowing users to discover the different services available from a single portal. A similar effort aiming at harmonizing the policies for gaining access to these infrastructures is under way.

At the same time, the e-IRG has recognized the need to provide European researchers with access to petaflop-range supercomputers in addition to high-throughput resources that are prevailing in EGI. PRACE aims to define the legal and organizational structures for a pan-European high-performance computing (HPC) service in the petaflop range. These petaflop-range systems are supposed to complement the existing European HPC e-infrastructure as pioneered by the DEISA project. DEISA is federating major European HPC centers in a common e-infrastructure providing seamless access to supercomputing resources and, thanks to a global shared file system, data stored at the various centers. This leads to a three-tier structure, with the European petaflop systems at Tier-0 being supported by leading national systems at Tier-1. Regional and midrange systems complement this HPC pyramid at Tier-2 as depicted in Figure 2.3.

Although the goals of PRACE/DEISA are similar to the ones of EGI/EGEE, the different usage and organizational requirements demand a different approach, and hence the establishment of two independent yet related infrastructures. For researchers, it is important, however, to have access to all infrastructures in a seamless manner; hence, a convergence of the services and operational models in a similar way as discussed in the EGI/NGI case mentioned earlier will be needed.

2.5.1 Layers of the Ecosystem

This convergence and the addition of other tools (like sensors, for instance) will eventually build the computing and data layer of the e-infrastructure ecosystem. Leveraging the physically wide area connectivity provided by the network infrastructure (operated by GéANT and the National Research and Education Networks in Europe), this computing and data layer facilitates the

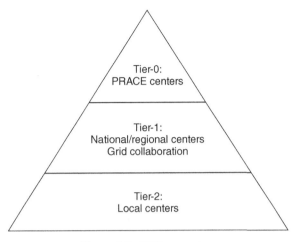

Figure 2.3 *HPC ecosystem.*

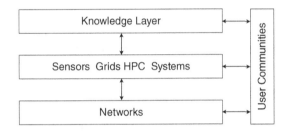

Figure 2.4 *e-Infrastructure ecosystem.*

construction of domain-specific knowledge layers that provide user communities with higher-level abstractions, allowing them to focus on their science rather than on the computing technicalities. The resulting three-layered ecosystem is depicted in Figure 2.4.

2.6 SUMMARY

In summary, a variety of different e-infrastructures are available today to support e-research. Convergence of these infrastructures in terms of interfaces and policies is needed to provide researchers with seamless access to the resources required for their research, independently of how the resource provisioning is actually managed. Eventually, a multilayer ecosystem will greatly reduce the need for scientists to manage their computing and data infrastructure, with a knowledge layer eventually providing high-level abstractions according to the needs of different disciplines. Initial elements of such an e-infrastructure ecosystem already exist, and Europe is actively striving for sustainability to ensure that it continues to build a reliable basis for e-research.

ACKNOWLEDGMENTS

The authors would like to thank all the members of EGEE and BalticGrid for their enthusiasm and excellent work, which made the EGEE and BalticGrid vision a reality. The authors would also like to thank the open source cloud developers—Eucalyptus and OpenNebula—making BalticCloud and BGi a reality.

REFERENCES

M Assuncao, A Costanzo, and R Buyya. Evaluating the cost-benefit of using cloud computing to extend the capacity of clusters. In *Proc. of 18th ACM Int'l Symposium on High Performance Distributed Computing*, pp. 141–150, 2009.

I. Foster and C. Kesselman, editors, *The Grid: Blueprint for A New Computing Infrastructure*. San Francisco, CA: Morgan Kaufmann, 2003.

I. Foster, C. Kesselman, and S. Tuecke. The anatomy of the grid. *The Int'l Journal of High Performance Computing Applications*, 15(3):200–222, 2001.

F. Gagliardi, B. Jones, and E. Laure. The EU DataGrid Project: Building and operating a large-scale grid infrastructure. In B. D. Martino, J. Dongarra, and A. Hoisie, editors, *Engineering the Grid: Status and Perspective*. Los Angeles, CA: American Scientific Publishers, 2006.

M. Lamanna and E. Laure, editors. The EGEE user forum experiences in using grid infrastructures. *Journal of Grid Computing*, 6(1), 2008.

E. Laure, S. Fisher, A. Frohner, et al. Programming the grid with gLite. *Computational Methods in Science and Technology*, 12(1):33–45, 2006.

E. Laure and B. Jones. Enabling Grids for E-sciencE—The EGEE project. In L. Wang, W. Jie, and J. Chen, editors, *Grid Computing: Infrastructure, Service, and Application*, Boca Raton, FL: CRC Press, 2009.

D. Nurmi, R. Wolski, C. Grzegorczyk, et al. The eucalyptus open-source cloud-computing system. In *Proc. of 9th IEEE/ACM International Symposium on Cluster Computing and the Grid*, pp. 124–131, 2009.

M. Riedel, E. Laure, T. Soddemann, et al. Interoperation of worldwide production e-science infrastructures. *Concurrency and Computation: Practice and Experience*, 21(8):961–990, 2009.

B. Sotomayor, R. Montero, I. M. Llorente, et al. Capacity leasing in cloud systems using the OpenNebula engine. In *Proc. of the Cloud Computing and its Applications 2008*, 2008.

Chapter *3*

Accelerated Many-Core GPU Computing for Physics and Astrophysics on Three Continents

Rainer Spurzem

National Astronomical Observatories of China, Chinese Academy of Sciences, Beijing, China, Astronomisches Rechen-Institut, Zentrum für Astronomie, University of Heidelberg, Heidelberg, Germany, and Kavli Institute for Astronomy and Astrophysics, Peking University, Beijing, China

Peter Berczik

National Astronomical Observatories of China, Chinese Academy of Sciences, Beijing, China, and Astronomisches Rechen-Institut, Zentrum für Astronomie, University of Heidelberg, Heidelberg, Germany

Ingo Berentzen

Zentrum für Astronomie, University of Heidelberg, Heidelberg, Germany

Wei Ge and Xiaowei Wang

Institute of Process Engineering, Chinese Academy of Sciences, Beijing, China

Hsi-Yu Schive

Department of Physics, National Taiwan University, Taibei, Taiwan

Keigo Nitadori

RIKEN AICS Institute, Kobe, Japan

Tsuyoshi Hamada

Nagasaki Advanced Computing Center, Nagasaki University, Nagasaki, Japan

José Fiestas

Astronomisches Rechen-Institut, Zentrum für Astronomie, University of Heidelberg, Heidelberg, Germany

Large-Scale Computing, First Edition. Edited by Werner Dubitzky, Krzysztof Kurowski, Bernhard Schott.
© 2012 John Wiley & Sons, Inc. Published 2012 by John Wiley & Sons, Inc.

3.1 INTRODUCTION

Theoretical numerical modeling has become a third pillar of sciences in addition to theory and experiment (in the case of astrophysics, the experiment is mostly substituted by observations). Numerical modeling allows one to compare theory with experimental or observational data in unprecedented detail, and it also provides theoretical insight into physical processes at work in complex systems. Similarly, data processing (e.g., of astrophysical observations) comprises the use of complex software pipelines to bring raw data into a form digestible for observational astronomers and ready for exchange and publication; these are, for example, mathematical transformations like Fourier analyses of time series or spatial structures, complex template analyses, or huge matrix–vector operations. Here, fast access to and transmission of data, too, require supercomputing capacities. However, sufficient resolution of multiscale physical processes still poses a formidable challenge, such as in the examples of few-body correlations in large astrophysical many-body systems or in the case of turbulence in physical and astrophysical flows.

We are undergoing a new revolution on parallel processor technologies, and a change in parallel programming paradigms, which may help to advance current software toward the exaflop scale and to better resolve and understand typical multiscale problems. The current revolution in parallel programming has been mostly catalyzed by the use of graphical processing units (GPUs) for general-purpose computing, but it is not clear whether this will remain the case in the future. GPUs have become widely used nowadays to accelerate a broad range of applications, including computational physics and astrophysics, image/video processing, engineering simulations, and quantum chemistry, just to name a few. GPUs are rapidly emerging as a powerful and cost-effective platform for high-performance parallel computing. The GPU Technology Conference 2010 held by NVIDIA in San Jose in autumn 2010[1] gave one snapshot of the breadth and depth of present-day GPU (super)computing applications. Recent GPUs, such as the NVIDIA Fermi C2050 Computing Processor, offer 448 processor cores and extremely fast on-chip-memory chip, as compared to only four to eight cores on a standard Intel or AMD central processing unit (CPU). Groups of cores have access to very fast shared memory pieces; a single Fermi C2050 device supports double-precision operations fully with a peak speed of 515 gigaflops; in this chapter, we also present results obtained from GPU clusters with previous generations of GPU accelerators, which have no (Tesla C870) or only very limited (Tesla C1060) double-precision support. We circumvented this by emulation of a few critical double-precision operations (Nitadori and Makino, 2008). More details can be found in the PhD thesis of one of us (K. Nitadori), "New Approaches to High-Performance *N*-Body Simulations with High-Order Integrator, New Parallel Algorithm, and Efficient Use of SIMD Hardware," University of Tokyo, in 2009.

[1] http://www.nvidia.com/gtc.

Scientists are using GPUs since more than 5 years already for scientific simulations, but only the invention of CUDA (Akeley et al., 2007; Hwu, 2011) as a high-level programming language for GPUs made their computing power available to any student or researcher with normal scientific programming skills. CUDA is presently limited to GPU devices of NVIDIA, but the open source language OpenCL will provide access to any type of many-core accelerator through an abstract programming language. Computational physics and astrophysics has been a pioneer in using GPUs for high-performance general-purpose computing (e.g., see the early AstroGPU workshop in Princeton 2007) through the information base.[2] Astrophysicists had an early start in the field through the Gravity Pipe (GRAPE) accelerator boards from Japan from 10 years ago (Makino et al., 2003; Fukushige et al., 2005). Clusters with accelerator hardware (GRAPE or GPU) have been used for gravitating many-body simulations to model the dynamics of galaxies and galactic nuclei with supermassive black holes (SMBHs) in galactic nuclei (Berczik et al., 2005, 2006; Berentzen et al., 2009; Just et al., 2011; Pasetto et al., 2011), in the dynamics of dense star clusters (Hamada and Iitaka, 2007; Portegies Zwart et al., 2007; Belleman et al., 2008), in gravitational lensing ray shooting problems (Thompson et al., 2010), in numerical hydrodynamics with adaptive mesh refinement (Wang and Abel, 2009; Schive et al., 2010; Wang et al., 2010a), in magnetohydrodynamics (Wong et al., 2009), or fast Fourier transform (FFT) (Cui et al., 2009). While it is relatively simple to obtain good performance with one or few GPUs relative to the CPU, a new taxonomy of parallel algorithms is needed for parallel clusters with many GPUs (Barsdell et al., 2010). Only "embarrassingly" parallel codes scale well even for large number of GPUs, while in other cases like hydrodynamics or FFT on GPU, the speedup is somewhat limited to 10–50 for the whole application, and this number needs to be carefully checked whether it compares the GPU performance with single or multicore CPUs. A careful study of the algorithms and their data flow and data patterns is useful and has led to significant improvements, for example, for particle-based simulations using smoothed particle hydrodynamics (SPH) (Berczik et al., 2007; Spurzem et al., 2009) or for FFT (Cui et al., 2009). Recently, new GPU implementations of fast-multipole methods (FMMs) have been presented and compared to Tree Codes (Yokota and Barba, 2010; Yokota et al., 2010). FMM codes have first been presented by Greengard and Rokhlin (1987). It is expected that on the path to exascale applications, further—possibly dramatic—changes in algorithms are required; at present, it is unclear whether the current paradigm of heterogeneous computing with a CPU and an accelerator device like GPU will remain dominant.

While the use of many-core accelerators is strongly growing in a large number of scientific and engineering fields, there are still only few codes able to fully harvest the computational power of parallel supercomputers with many GPU devices, as they have recently become operational in particular

---

[2] http://www.astrogpu.org.

(but not restricted to) in China. In China, GPU computing is blooming; the top and third spots in the list of 500 fastest supercomputers in the world[3] are now occupied by Chinese GPU clusters, and one of the GPU clusters used for results in this chapter is on rank number 28 (Mole-8.5 computer; see next discussion and Wang et al., 2010b). In this chapter, we present in some detail an astrophysical N-body application for star clusters and galactic nuclei and star clusters, which is currently our mostly well-tested and also heavily used application. Furthermore, somewhat less detailed, we present other applications scaling equally very well, such as an adaptive mesh refinement hydrodynamic code, using (among other parts) an FFT and relaxation methods to solve Poisson's equation, and give some overview on physical and process engineering simulations.

3.2 ASTROPHYSICAL APPLICATION FOR STAR CLUSTERS AND GALACTIC NUCLEI

Dynamical modeling of dense star clusters with and without massive black holes poses extraordinary physical and numerical challenges; one of them is that gravity cannot be shielded, such as electromagnetic forces in plasmas; therefore, long-range interactions go across the entire system and couple nonlinearly with small scales; high-order integration schemes and direct force computations for large numbers of particles have to be used to properly resolve all physical processes in the system. On small scales, correlations inevitably form already early during the process of star formation in a molecular cloud. Such systems are dynamically extremely rich; they exhibit a strong sensitivity to initial conditions and regions of phase space with deterministic chaos.

After merging two galaxies in the course of cosmological structure formation, we start our simulations with two SMBHs embedded in a dense star cluster, separated by some 1000 pc (1 pc, 1 parsec, about 3.26 light years, or $3.0857 \cdot 10^{18}$ cm). This is a typical separation still accessible to astrophysical observations (Komossa et al., 2003). Nearly every galaxy harbors an SMBH, and galaxies build up from small to large ones in a hierarchical manner through mergers following close gravitational encounters. However, the number of binary black holes observed is relatively small, so there should be some mechanism by which they get close enough to each other to coalesce under emission of gravitational wave emission. Direct numerical simulations of Einstein's field equations start usually at a black hole separation in the order of 10–50 Schwarzschild radii, which is, for an example of a 1 million solar mass black hole (similar to the one in our own galactic center), about 10^{-5} pc. Therefore, in order to obtain a merger, about eight orders of magnitude in separation need to be bridged. In our recent models, we follow in one coherent

---

[3] http://www.top500.org.

direct N-body simulation how interactions with stars of a surrounding nuclear star cluster, combined with the onset of relativistic effects, lead to a black hole coalescence in galactic nuclei after an astrophysically modest time of order 10^8 years (Berentzen et al., 2009; Preto et al., 2011). Corresponding to the multiscale nature of the problem in space, we have a large range of timescales to be covered accurately and efficiently in the simulation. Orbital times of SMBHs in galactic nuclei after galactic mergers are of the order of several million years; in the interaction phase with single stars, the orbital time of a gravitationally bound supermassive binary black hole goes down to some 100 years— at this moment, there is a first chance to detect its gravitational wave emission through their influence on pulsar timing (Lee et al., 2011). Energy loss due to Newtonian interactions with field stars interplays with energy loss due to gravitational radiation emission; the latter becomes dominant in the final phase (at smaller separations) when the black hole binary enters the wave band of the planned laser interferometer space antenna (LISA[4]), where one reaches 0.01-Hz orbital frequency. Similarly, in a globular star cluster, timescales can vary between a million years (for an orbit time in the cluster) to hours (orbital time of the most compact binaries). The nature of gravity favors such strong structuring properties since there is no global dynamic equilibrium. Gravitationally bound subsystems (binaries) tend to exchange energy with the surrounding stellar system in a way that increases their binding energy, thus moving further away from a global equilibrium state. This behavior can be understood in terms of self-gravitating gas spheres undergoing gravothermal catastrophe (Lynden-Bell and Wood, 1968), but it occurs in real star clusters on all scales. Such kind of stellar systems, sometimes called dense or gravothermal stellar systems, demands special high-accuracy integrators due to secular instability, deterministic chaos, and strong multiscale behavior. Direct high-order N-body integrators for this type of astrophysical problem have been developed by Aarseth (see for reference Aarseth, 1999b, 2003). They employ fourth-order time integration using a Hermite scheme, hierarchically blocked individual particle time steps, an Ahmad–Cohen neighbor scheme, and regularization of close, few-body systems.

Direct N-Body Codes in astrophysical applications for galactic nuclei, galactic dynamics, and star cluster dynamics usually have a kernel in which direct particle–particle forces are evaluated. Gravity as a monopole force cannot be shielded on large distances, so astrophysical structures develop high-density contrasts. High-density regions created by gravitational collapse coexist with low-density fields, as is known from structure formation in the universe or the turbulent structure of the interstellar medium. A high-order time integrator in connection with individual, hierarchically blocked time steps for particles in a direct N-body simulation provides the best compromise between accuracy, efficiency, and scalability (Makino and Hut, 1988; Spurzem, 1999; Aarseth, 1999a, 1999b; Harfst et al., 2007). With GPU hardware, up to a few million

---

[4] http://lisa.nasa.gov/.

bodies could be reached for our models (Berczik et al., 2005, 2006; Gualandris and Merritt, 2008). Note that while Greengard and Rokhlin (1987) already mentioned that their algorithm can be used to compute gravitational forces between particles to high accuracy, Makino and Hut (1988) found that the self-adaptive hierarchical time-step structure inherited from Aarseth's codes improves the performance for spatially structured systems by $\mathcal{O}(N)$it means that, at least, for astrophysical applications with high-density contrast, FMM is not a priori more efficient than direct N-body (which sometimes is called "brute force," but that should only be used if a shared time step is used, which is not the case in our codes). One could explain this result by comparing the efficient spatial decomposition of forces (in FMM, using a simple shared time step) with the equally efficient temporal decomposition (in direct N-body, using a simple spatial force calculation).

On the other hand, cosmological N-body simulations use thousand times more particles (billions, order of 10^9), at the price of allowing less accuracy for the gravitational force evaluations, either through the use of a hierarchical decomposition of particle forces in time (Ahmad and Cohen, 1973; Makino and Aarseth, 1992; Aarseth, 2003) or in space (Barnes and Hut, 1986; Makino, 2004; Springel, 2005). Another possibility is the use of fast-multipole algorithms (Greengard and Rokhlin, 1987; Dehnen, 2000, 2002; Yokota and Barba, 2010; Yokota et al., 2010) or particle–mesh (PM) schemes (Hockney and Eastwood, 1988; Fellhauer et al., 2000), which use FFT for their Poisson solver. PM schemes are the fastest for large systems, but their resolution is limited to the grid cell size. Adaptive codes use direct particle–particle forces for close interactions below grid resolution (Couchman et al., 1995; Pearce and Couchman, 1997). But for astrophysical systems with high-density contrasts, tree codes are more efficient. Recent codes for massively parallel supercomputers try to provide adaptive schemes using both tree and PM, such as the well-known GADGET and TreePM codes (Xu, 1995; Springel, 2005; Yoshikawa and Fukushige, 2005; Ishiyama et al., 2009).

3.3 HARDWARE

We present results obtained from GPU clusters using NVIDIA Tesla C1060 cards in Beijing, China (Laohu cluster with 85 Dual Intel Xeon nodes and with 170 GPUs); NVIDIA Fermi C2050 cards also in Beijing, China (Mole-8.5 cluster with 372 dual Xeon nodes, most of which have 6 GPUs, delivering in total 2000 Fermi Tesla C2050 GPUs); in Heidelberg, Germany, using NVIDIA Tesla C870 (pre-Fermi single-precision-only generation) cards (Kolob cluster with 40 Dual Intel Xeon nodes and with 40 GPUs.); and Berkeley at NERSC/LBNL using again the NVIDIA Fermi Tesla C2050 cards (Dirac cluster with 40 GPUs) (Fig. 3.1).

In Germany, at Heidelberg University, our teams have operated a many-core accelerated cluster using the GRAPE hardware for many years (Spurzem

Figure 3.1   *Left: NAOC GPU cluster in Beijing, 85 nodes with 170 NVIDIA Tesla C1060 GPUs, 170-teraflop hardware peak speed, installed in 2010. Right: Frontier Kolob cluster at ZITI Mannheim, 40 nodes with 40 NVIDIA Tesla C870 GPU accelerators, 17-teraflop hardware peak speed, installed in 2008.*

et al., 2004, 2007, 2008; Harfst et al., 2007). We have in the meantime migrated from GRAPE to GPU (and also partly FPGA) clusters (Spurzem et al., 2009, 2010, 2011), and part of our team is now based at the National Astronomical Observatories of China (NAOC) of the Chinese Academy of Sciences (CAS) in Beijing. NAOC is part of a GPU cluster network covering 10 institutions of CAS, aiming for high-performance scientific applications in a cross-disciplinary way. The top-level cluster in this network is the recently installed Mole-8.5 cluster at the Institute of Process Engineering of the Chinese Academy of Sciences (IPE/CAS) in Beijing (2-petaflop single-precision peak), from which we also show some preliminary benchmarks. The entire CAS GPU cluster network has a total capacity of nearly 5-petaflop single-precision peak. In China, GPU computing is blooming; the top and third spots in the list of 500 fastest supercomputers in the world[5] are now occupied by Chinese GPU clusters. The top system in the CAS GPU cluster network is currently number 28 (Mole-8.5 at IPE). Research and teaching in CAS institutions is focused on broadening the computational science base to use the clusters for supercomputing in basic and applied sciences.

3.4 SOFTWARE

The test code that we use for benchmarking on our clusters is a direct *N*-body simulation code for astrophysics, using a high-order Hermite integration scheme and individual block time steps (the code supports time integration of

---

[5] http://www.top500.org.

particle orbits with fourth-, sixth, and eighth-order schemes). The code is called φGPU; it has been developed from our earlier published versions φGRAPE (Harfst et al., 2007). It is parallelized using Message Passing Interface (MPI) and, on each node, using many cores of the special hardware. The code was mainly developed and tested by K. Nitadori and P. Berczik (see also Hamada and Iitaka, 2007) and is based on an earlier version for GRAPE clusters (Harfst et al., 2007). The code is written in C++ and is based on Nitadori and Makino (2008) an earlier CPU serial code (yebisu).

We used and tested the present version of the φGPU code only with the recent GNU compilers (version 4.x). For all results shown here, we used a Plummer-type gravitational potential softening, $\varepsilon = 10^{-4}$, in units of the virial radius. More details will be published in an upcoming publication (Berczik et al., 2011).

The MPI parallelization was carried out in the same "j" particle parallelization mode as in the earlier φGRAPE code (Harfst et al., 2007). The particles were divided equally between the working nodes and in each node, we calculated only the fractional forces for the active "i" particles at the current time step. Due to the hierarchical block time-step scheme, the number N_{act} of active particles (due for a new force computation at a given time level) is usually small compared to the total particle number N, but its actual value can vary from $1 \ldots N$. We obtained the full forces from all the particles acting on the active particles after using the global MPI_SUM communication routines.

We used native GPU support and direct code access to the GPU with only CUDA. Recently, we used the latest CUDA 3.2 (but the code was developed and working also with the "older" CUDA compilers and libraries). Multi-GPU support is achieved through MPI parallelization; each MPI process uses only a single GPU, but we can start two MPI processes per node (to use effectively, e.g., the dual quad core CPUs and the two GPUs in the NAOC cluster) and in this case each, MPI process uses its own GPU inside the node. Communication always (even for the processes inside one node) works via MPI. We do not use any of the possible OMP (multithread) features of recent gcc 4.x compilers inside one node.

3.5 RESULTS OF BENCHMARKS

Figures 3.2 and 3.3 show results of our benchmarks. In the case of Laohu, we use maximum 164 GPU cards (three nodes; i.e., six cards were down during the test period). Here, the largest performance was reached for 6 million particles, with 51.2 teraflops in total sustained speed for our application code, in an astrophysical run of a Plummer star cluster model, simulating one physical time unit (about one-third of the orbital time at the half-mass radius). Based on these results, we see that we get a sustained speed for 1 NVIDIA Tesla C1060 GPU card of 360 gigaflops (i.e., about one-third of the theoretical hardware peak speed of 1 teraflop). Equivalently, for the smaller and older Kolob

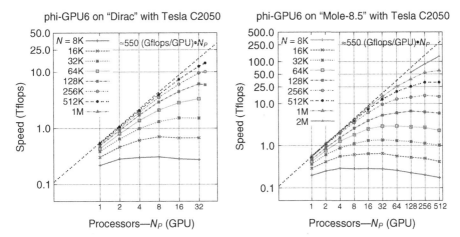

Figure 3.2 *Strong scaling for different problem sizes. Left: Dirac Fermi Tesla C2050 GPU system at NERSC/LBNL, almost 18 teraflops reached for 1 million particles on 40 GPUs. Each line corresponds to a different problem size (particle number), which is given in the key. Note that the linear curve corresponds to ideal scaling. Right: same benchmark simulations, but for the Mole-8.5 GPU cluster at IPE in Beijing, using up to 512 Fermi Tesla C2050 GPUs, reaching a sustained speed of 130 teraflops (for 2 million particles). If one would use all 2000 GPUs on the system, a sustained speed of more than 0.4 petaflop is feasible. This is the subject of ongoing present and future work.*

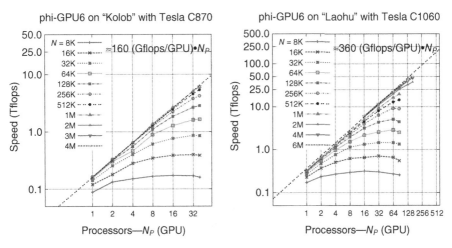

Figure 3.3 *Left: Same benchmark simulations as in Figure 3.2, but for the Frontier Kolob cluster with Tesla C870 GPUs at University of Heidelberg; 6.5 teraflops reached for 4 million particles on 40 GPUs. Right: NAOC GPU cluster in Beijing; speed in teraflops reached as a function of the number of processes, each process with one GPU; 51.2 teraflops sustained were reached with 164 GPUs (three nodes with 6 GPUs were down at the time of testing).*

cluster with 40 NVIDIA Tesla C870 GPUs in Germany, we obtain 6.5 teraflops (with 4 million particles). This is 160 gigaflops per card.

On the new clusters Dirac and Mole-8.5, where we used the NVIDIA Fermi Tesla C2050 cards, we get the maximum performance of 550 gigaflops per card. We achieve the absolute record in the performance on a Mole-8.5 cluster when we run our test simulation (even for a relatively "low" particle number—2 million) on 512 nodes and get over the 130-teraflop total performance. In principle, for a larger particle number (in the order of 10 million), we see that the maximum performance that we can get on the whole cluster (on ≈ 2000 GPUs) is around 0.4 petaflop.

We have presented exemplary implementations of direct gravitating N-body simulations and an adaptive mesh hydrodynamics code with self-gravity (Schive et al., 2010) using large GPU clusters in China and elsewhere. The overall parallelization efficiency of our codes is very good. It is about 30% of the GPU peak speed in Figure 3.2 for the embarrassingly parallel direct N-body code and still significant (in the order of 20–40 speedups for each GPU) for adaptive mesh hydrodynamic simulations. The larger N-body simulations (several million particles) show nearly ideal strong scaling (linear relation between speed and number of GPUs) up to our present maximum number of nearly 170 GPUs—no strong sign of a turnover yet due to communication or other latencies. Therefore, we are currently testing the code implementation on much larger GPU clusters, such as the Mole-8.5 of IPE/CAS.

The wall clock time T needed for our particle-based algorithm to advance the simulation by a certain physical time (usually one crossing time units) integration interval scales as

$$T = T_{host} + T_{GPU} + T_{comm} + T_{MPI}, \tag{3.1}$$

where the components of T are (from left to right) the computing time spent on the host, on the GPU, the communication time to send data between the host and GPU, and the communication time for MPI data exchange between the nodes. In our present implementation, all components are blocking, so there is no hiding of communication. This will be improved in further code versions, but for now, it eases profiling (Table 3.1).

In the case of the φGPU code (as in the other direct NBODY codes discussed next), we used the blocked hierarchical individual time-step scheme (HITS) and a Hermite high-order time integration scheme of at least the fourth order for the integration of the equation of motions for all particles (Makino and Aarseth, 1992). In the case of HITS in every individual time steps, we integrate the motion only for s particles, a number that is usually much less compared to the total number of particles N. Its average value $\langle s \rangle$ depends on the details of the algorithm and on the particle configuration integrated. According to a simple theoretical estimate, it is $\langle s \rangle \propto N^{2/3}$ (Makino, 1991), but the real value of the exponent deviates from two-thirds, depending on the initial model and details of the time-step choice (Makino and Hut, 1988).

TABLE 3.1   Properties of the Fermi GPU Cluster at the Institute of Process Engineering of the Chinese Academy of Sciences (IPE/CAS)

Data of the Mole-8.5 System

Item	Quantity
Peak performance single precision	2 petaflops
Peak performance double precision	1 petaflop
Linpack sustained performance	207.3 teraflops
Megaflops per watt	431
Number of nodes/number of GPUs (type)	372/2000 (Fermi Tesla C2050)
Total memory RAM	17.8 terabytes
Total memory VRAM	6.5 terabytes
Total hard disk	720 terabytes
Management communication	H3C Gigabit Ethernet
Message passing communication	Mellanox InfiniBand Quad Data Rate
Occupied area	150 *m*2
Weight	12.6 tons
Max power	600 kW (computing)
	200 kW (cooling)
Operating system	CentOS 5.4, PBS
Monitor	Ganglia, GPU monitoring
Languages	C, C++;, CUDA

This system is the largest GPU cluster in Beijing, the third Chinese cluster, with rank 28 in the worldwide Top500 list (as of November 2010). It has been used for some of our *N*-body benchmarks, especially for the timing model, and by the physics simulations at IPE. Note that it has a relatively large number of GPUs per node, but our communication performance was not significantly affected (see comparison plots with Dirac cluster in Berkeley, which has only one GPU per node).

We use a detailed timing model for the determination of the wall clock time needed for different components of our code on CPU and GPU, which is then fitted to the measured timing data. Its full definition is given in Table 3.2.

In practice, we see that only three terms play any relevant role to understand the strong and weak scaling behavior of our code. These are the force computation time (on GPU) T_{GPU} and the message passing communication time T_{MPI}, within which we can distinguish a bandwidth-dependent part (scaling as $s \log(N_{GPU})$) and a latency-dependent part (scaling as $\tau_{lat} \log(N_{GPU})$); the latency is only relevant for a downturn of efficiency for strong scaling at a relatively large N_{GPU}. Starting in the strong scaling curves from the dominant term at a small N_{GPU}, there is a linearly rising part, as one can see, in all curves in Figures 3.2–3.4 (most clearly in Fig. 3.4), which corresponds only to the force computation on GPU, while the turnover to a flat curve is dominated by the MPI communication time between the computing nodes, T_{MPI}.

To find a model for our measurements, we use the ansatz,

$$P = (\text{total number of floating point operations})/T, \qquad (3.2)$$

where T is the computational wall clock time needed. For one block step, the total number of floating point operations is $\gamma\langle s\rangle N$, where γ defines how many

TABLE 3.2 Breaking Down the Computational Tasks in A Parallel Direct *N*-Body Code with Individual Hierarchical Block Time Steps

Components in Our Timing Model for Direct *N*-Body Code

Task	Expected Scaling	Timing Variable
Active particle determination	$\mathcal{O}(s \log[s])$	T_{host}
All particle prediction	$\mathcal{O}(N/N_{GPU})$	T_{host}
"j" part. send. to GPU	$\mathcal{O}(N/N_{GPU})$	T_{comm}
"i" part. send. to GPU	$\mathcal{O}(s)$	T_{comm}
Force computation on GPU	$\mathcal{O}(Ns/N_{GPU})$	T_{GPU}
Receive the force from GPU	$\mathcal{O}(s)$	T_{comm}
MPI global communication	$\mathcal{O}((\tau_{lat} +; s)\log(N_{GPU}))$	T_{MPI}
Correction/advancing "i" particle	$\mathcal{O}(s)$	T_{host}

At every block time-step level, we denote $s \leq N$ particles, which should be advanced by the high-order corrector as active or "i" particles, while the field particles, which exert forces on the i particles to be computed, are denoted as "j" particles. Note that the number of j particles in our present code is always N (full particle number), but in more advanced codes like NBODY6 discussed next, the Ahmad–Cohen neighbor scheme uses a much smaller number of j particles for the more frequent neighbor force calculation. We also have timing components for low-order prediction of all j particles and distinguish communication of data from host to GPU and return, and through the MPI message passing network.

Figure 3.4 *Left: The Mole-8.5 Cluster at the Institute of Process Engineering in Beijing. It consists of 372 nodes, most with 6 Fermi Tesla C2050 GPUs. Right: single node of Mole-8.5 system (courtesy of IPE, photos by Xianfeng He).*

floating point operations our particular Hermite scheme requires per particle per step, and we have

$$P_s = \frac{\gamma N \langle s \rangle}{T_s} = \frac{\gamma N \langle s \rangle}{\alpha N \langle s \rangle / N_{gpu} + \beta (\tau_{lat} + \langle s \rangle) \log (N_{gpu})}, \qquad (3.3)$$

where T_s is the computing time needed for one average block step in time (advancing $\langle s \rangle$ particles). The reader with interest in more detail how this formula can be theoretically derived for general-purpose parallel computers is referred to Dorband et al. (2003). α, β, and τ_{lat} are hardware time constants for the floating point calculation on GPU, for the bandwidth of the interconnect hardware used for message passing and its latency, respectively.

Our timing measurements are done for an integration over one physical time unit in normalized units ($t = 1$, which is equivalent to approximately one-third of a particle's orbital crossing time at the half-mass radius), so it is more convenient to multiply the numerator and denominator of Equation 3.3 with the average number $\langle n \rangle$ of steps required for an integration over a physical timescale t; it is $\langle n \rangle \propto t / \langle dt \rangle$, where $\langle dt \rangle$ is the average individual time step. In a simple theoretical model, our code should asymptotically scale with N^2, so we would expect $N \langle s \rangle \langle n \rangle \propto N^2$. However, our measurements deliver a slightly less favorable number $\langle s \rangle \langle n \rangle \propto N^{1+x}$, with $x = 0.31$, a value in accord with the results of Makino and Hut (1988). Hence, we get for the integration over one time unit

$$P_s \approx \frac{\gamma N^{2+x}}{\alpha N^{2+x} / N_{gpu} + \beta (\tau_{lat} + N^{1+x}) \log (N_{gpu})}. \qquad (3.4)$$

The parameter $x = 0.31$ is a particular result for our case of the sixth-order HITS and the particular initial model used for the N-body system, Plummer's model as in Makino and Hut (1988). x is empirically determined from our timing measurements as shown in Figure 3.5. The parameters α, β, γ, and τ_{lat} can be determined as well for each particular hardware used. The timing formula can then be used to approximate our code calculation "speed" for any other number of particles, GPUs, or different hardware parameters. For example, on the Mole-8.5 system, we see that for $N = 10\,m$ particles, if we are using 2000 GPU cards on the system, we expect to get ≈ 390 teraflops (compare with Fig. 3.5). If we use our scaling formula for the much higher node-to-node bandwidth of the Tianhe-1 system at Tianjin Supercomputing Center (this is the number one supercomputer according to the Top500 list of November 2010, with 7000 NVIDIA Fermi Tesla GPUs and 160 gigabit/s node-to-node bandwidth), we can possibly reach a sustained performance on the order of petaflops. This is the subject of future research.

To our knowledge, the direct N-body simulation with 6 million bodies in the framework of a so-called Aarseth style code (Hermite scheme, sixth order, hierarchical block time step, integrating an astrophysically relevant Plummer

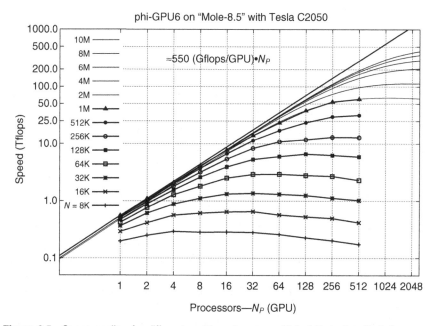

Figure 3.5 *Strong scaling for different problem sizes on a Mole-8.5 cluster. Each line corresponds to a different problem size (particle number), which is given in the key. The sequence of lines in the plot corresponds to the sequence of lines in the key (from top to bottom). Thicker lines with dots or symbols are obtained from our timing measurements. Thinner lines show the extrapolation for larger N_{GPU} and for larger N according to our timing model. As one can see, we reach 550 gigaflops per GPU card, in total on 512 GPUs about 280-teraflop sustained code performance for our code. An extrapolation to 2000 GPUs shows we can reach 390 teraflops on Mole-8.5 for 10 million particles.*

model with a core–halo structure in density for a certain physical time) is the largest of such simulations that exists so far. However, the presently used parallel MPI-CUDA GPU code φGPU is on the algorithmic level of NBODY1 (Aarseth, 1999b); though it is already strongly used in production, useful features such as regularization of few-body encounters and an Ahmad–Cohen neighbor scheme (Ahmad and Cohen, 1973) are not yet implemented. Only with those the code would be equivalent to NBODY6, which is the most efficient code for single workstations (Aarseth, 1999b, 2003), eventually with acceleration on a single node by one or two GPUs (work by Aarseth & Nitadori; see NBODY6[6]). NBODY6++ (Spurzem, 1999) is a massively parallel code corresponding to NBODY6 for general-purpose parallel computers. An NBODY6++ variant using many GPUs in a cluster is work in progress. Such a code could potentially reach the same physical integration time (with same

---

[6] http://www.ast.cam.ac.uk/~sverre/web/pages/nbody.htm.

accuracy) using only one order of magnitude less floating point operations. The NBODY6 codes are algorithmically more efficient than φGPU or NBODY1 because they use an Ahmad–Cohen neighbor scheme (Ahmad and Cohen, 1973), which reduces the total number of full force calculations needed again (in addition to the individual hierarchical time-step scheme); that is, the proportionality factor in front of the asymptotic complexity N^2 is further reduced.

We have shown that our GPU clusters for the very favorable direct N-body application reach about one-third of the theoretical peak speed sustained for a real application code with individual block time steps. In the future, we will use larger Fermi-based GPU clusters such as the Mole-8.5 cluster at the IPE/CAS in Beijing and more efficient variants of our direct N-body algorithms; details of benchmarks and science results, and the requirements to reach exascale performance, will be published elsewhere.

3.6 ADAPTIVE MESH REFINEMENT HYDROSIMULATIONS

The team at National Taiwan University has developed an adaptive mesh refinement (AMR) code named GAMER to solve astrophysical hydrodynamic problems (Schive et al., 2010). The AMR implementation is based on constructing a hierarchy of grid patches with an oct-tree data structure. The code adopts a hybrid CPU/GPU model in which both hydrodynamic and gravity solvers are implemented into the GPU and the AMR data structure is manipulated by the CPU. For strong scaling, considerable speedup is demonstrated for up to 128 GPUs, with excellent performance shown in Figures 3.6 and 3.7.

More recently, the GAMER code is further optimized for supporting several directionally unsplit hydrodynamic schemes and the OpenMP parallelization (Schive et al., accepted). By integrating hybrid MPI/OpenMP parallelization with GPU computing, the code can fully exploit the computing power in a heterogeneous CPU/GPU system. The Figure 3.8 shows the performance benchmark on the Dirac cluster at NERSC/LBNL. The maximum speedups achieved in the 32-GPU run are 71.4 and 18.3 as compared to the CPU-only 32-single-core and 32-quad-core performances, respectively. Note that the 32-GPU speedup drops about 12% mainly due to the MPI communication and the relatively lower spatial resolution (and hence higher surface/volume ratio) compared to that of the benchmark performed on the Beijing Laohu cluster. This issue can be alleviated by increasing the spatial resolution and also by overlapping communication with computation.

3.7 PHYSICAL MULTISCALE DISCRETE SIMULATION AT IPE

Discrete simulation is, in a sense, more fundamental and straightforward as compared with other numerical methods based on continuum models since

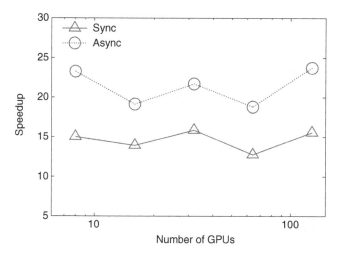

Figure 3.6 *Performance speedup of the GAMER code as a function of the number of GPUs, measured on the Beijing Laohu cluster. The test problem is a purely baryonic cosmological simulation of ΛCDM, in which the root-level resolution is 256^3 and seven refinement levels are used, giving $32,768^3$ effective resolutions. For each data point, we compare the performance by using the same number of GPUs and CPU cores. The circles and triangles show the timing results with and without the concurrent execution between CPUs and GPUs, respectively. The maximum speedup achieved in the 128-GPU run is about 24.*

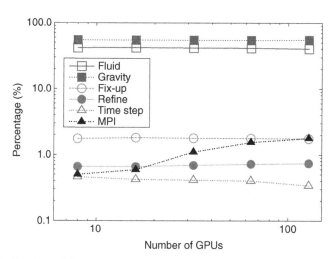

Figure 3.7 *Fractions of time consumed in different parts of the GAMER code. This includes the hydrodynamic solver (open squares), gravity solver (filled squares), coarse-grid data correction (open circles), grid refinement (filled circles), computing time step (open triangles), and MPI communication (filled triangles). It shows that the MPI communication time even for large number of GPUs uses only a small percentage of time (order of 1%), and hence the code will be able to scale well to even much larger GPU numbers.*

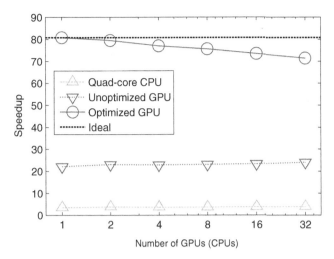

Figure 3.8 *Performance speedup of the latest GAMER code of 2011, measured on the Dirac cluster at NERSC/LBNL. The root-level resolution is 256^3 and only four refinement levels are used. The GPU performance is compared to that of CPU runs without OpenMP and GPU acceleration. Several optimizations are implemented in the fully optimized code, including the asynchronous memory copy, the concurrent execution between CPU and GPU, and the OpenMP parallelization. The quad-core CPU performance is also shown for comparison.*

the world is naturally composed of particles at very small and large scales, such as fundamental particles, atoms, and molecules on one hand and stars and galaxies on the other hand. However, continuum methods are traditionally considered more efficient as each element in these methods presents a statistically enough number of particles. This faith has changed in recent years with the dramatic development of parallel computing. It turns out that although the peak performance of (parallel) supercomputers is increasing at a speed higher than Moore's law, the sustainable performance of most numerical softwares is far behind it, sometimes only several percent of it, and the percentage decreases with system scale inevitably. The complex data dependence and hence communication overheads inherent for most continuum-based numerical methods present a major cause of this inefficiency and poor scalability. In comparison, discrete simulation methods, such as molecular dynamic (MD) simulations, dissipative particle dynamics (DPD), lattice Boltzmann method (LBM), discrete particle methods (DEM) and SPH, and so on, heavily rely on local interactions, and their algorithms are inherently parallel. In the final analysis, this is rooted in the physical parallelism of the physical model behind these methods. It is worthy to mention that coarse-grained particles such DPD and PPM (Ge and Li, 2003a) are now capable of simulating apparently continuous systems at a computational cost fairly comparable to continuum methods, and macroscale particle methods such SPH and MaPPM (Ge and Li,

2001, 2003b) can also be understood as a special kind of numerical discretizing of continuum models.

In recent years, with the flourish of many-core computing technology, such as the use of GPUs (graphic processing unit) for scientific and engineering computing, this virtue of discrete methods is best demonstrated and further explored. A general model for many-core computing of discrete methods is "divide and conquer." A naive implementation is to decompose the computed domain into many subdomains, which are then assigned to different processors for parallel computing of particle–particle interactions and movements. The assignment changes as the physical location of transfer from one subdomain to another. Communications, therefore, only occur at neighboring subdomains. Most practical implementations, however, use more advanced techniques, such as dynamic load balance and monotonic Lagrangian grid (Lambrakos and Boris, 1987), to minimize the waiting and communication among different processors. Within each processor, each pair of particle–particle interactions and each particle-state updating are also parallel in principle, which can be carried out by each core of the processors. Currently, most many-core processes like GPUs are still working as an external device to the CPU, so data copy between the main memory and the device memory is still necessary, and the communication between many-core processors across different computing nodes is routed by CPUs. A combined CPU–GPU computing mode is under development, which may further reduce this communication overhead.

Some of the discrete simulation work carried out at IPE/CAS using GPUs has been introduced in a Chinese monograph (Chen et al., 2009), and in some recent publications, they have covered MD simulation of multiphase micro- and nanoflow (Chen et al., 2008), polymer crystallization (Xu et al., 2009) and silicon crystal, computational fluid dynamics (CFD) simulation of cavity flow (Li et al., 2009) and gas–solid suspension, and so on. All the simulations introduced earlier have been carried out on the multiscale high-performance computing (HPC) systems established at IPE. The first system, Mole-9.7, put into use on February 18, 2008, consists of 120 HP xw8600 workstations, each installed with two NVIDIA Tesla C870 GPGPU cards and two Intel Xeon 5430 CPUs, and reached a peak performance of 120 teraflops in a single precision. The system is connected by an all-to-all switch together with a 2-D torus topology of Gigabit Ethernet, which speeds up adjacent communication dominated in discrete simulations. Its successor, Mole-8.7, is announced on April 20, 2009 as the first supercomputer of China with 1.0-petaflop peak performance in a single precision (Chen et al., 2009). Both NVIDIA and AMD GPU are integrated in this system. The designing philosophy is the consistency among hardware, software, and the problems to be solved, based on the multiscale method and discrete simulation approaches developed at IPE. The system has nearly 400 nodes connected by Gigabit Ethernet and DDR InfiniBand network.

Then, in 2010, IPE built the new system, Mole-8.5, which is the first GPU cluster using Fermi in the world. With the powerful computational resource of Mole-8.5 and the multiscale software developed by IPE, several large-scale applications have been successfully run on Mole-8.5:

- An MD simulation of dynamic structure of a whole H1N1 influenza virion in solution is simulated at the atomic level for the first time. The simulation system includes totally 300 million atoms in a periodic cube with an edge length of 148.5 nm. Using 288 nodes with 1728 Fermi Tesla C2050, the simulation proceeds at 770 ps/day with an integration time step of 1fs (Xu et al., 2010b).

- A quasi-real-time DEM simulation of an industrial rotating drum, the size of which is 13.5 m long by 1.5 m in diameter, is performed. The simulation system contains about 9.6 million particles. Nearly 1/11 real speed is achieved using 270 GPUs together with online visualization (Xu et al., 2010a).

- Large-scale direct numerical simulations of gas–solid fluidization have been carried out, with systems of about 1 million solid particles and 1 billion fluid particles in 2-D using 576 GPUs, and of about 100,000 solid particle and 0.4 billion fluid particles in 3-D using 224 GPUs. The largest system we have run utilized 1728 GPUs with an estimated performance of 33 teraflops in double precision (Xiong et al., 2010).

- A large-scale parallel MD simulation of single-crystalline silicon nanowire containing about 1.5 billion silicon atoms with many-body potential is conducted using 1500-GPU cards with a performance of about 227 teraflops in single precision (Hou and Ge, 2011).

3.8 DISCUSSION AND CONCLUSIONS

We have presented exemplary implementations of parallel codes using many GPUs as accelerators, so combining message passing parallelization with many-core parallelization, and have discussed their benchmarks using up to 512 Fermi Tesla GPUs in parallel, mostly on the Mole-8.5 hardware of the IPE/CAS in Beijing, but also on the Laohu Tesla C1070 cluster of the National Astronomical Observatories of CAS in Beijing and smaller clusters in Germany and in the United States. For direct high-accuracy gravitating N-body simulations, we discussed how self-gravity, because it cannot be shielded, generates inevitably strong multiscale structures in space and time, spanning many orders of magnitude. This requires special codes, which nevertheless scale with a high efficiency on GPU clusters. Also, we present an adaptive mesh hydrodynamic code including a gravity solver using FFT and relaxation methods and physical algorithms used for multiscale flows with particles. So, our codes are examples that it is possible to reach the subpetaflop scale in sustained speed for realistic application software with large GPU clusters. Whether our programming models can be scaled up for future hardware and the exaflop scale, however, remains yet to be studied.

ACKNOWLEDGMENTS

CAS has supported this work by a Visiting Professorship for Senior International Scientists, Grant Number 2009S1-5 (RS), and NAOC/CAS through the Silk Road Project (RS, PB, JF partly). IPE/CAS and the High Performance Computing Center at NAOC/CAS acknowledge financial support by the Ministry of Finance under the grant ZDYZ2008-2 for the supercomputers Mole-8.5 and Laohu, used for simulations of this chapter. RS and PB want to thank Xue Suijian for valuable advice and support. We thank the computer system support team at NAOC (Gao Wei and Cui Chenzhou) for their support to run the Laohu cluster.

We gratefully acknowledge computing time on the Dirac cluster of NERSC/LBNL in Berkeley and thank Hemant Shukla, John Shalf, and Horst Simon for providing the access to this cluster and for cooperation in the International Center of Computational Science,[7] as well as the helpful cooperation of Guillermo Marcus, Andreas Kugel, Reinhard Männer, Robi Banerjee, and Ralf Klessen in the GRACE and Frontier Projects at the University of Heidelberg (at ZITI and ITA/ZAH).

Simulations were also performed on the GRACE supercomputer (grants I/80 041-043 and I/81 396 of the Volkswagen Foundation and 823.219-439/30 and /36 of the Ministry of Science, Research and the Arts of Baden-Württemberg) and the Kolob cluster funded by the Frontier Project at the University of Heidelberg. PB acknowledges the special support by the NAS Ukraine under the Main Astronomical Observatory GRAPE/GRID computing cluster project.[8] PB's studies are also partially supported by the program Cosmomicrophysics of NAS Ukraine. The Kolob cluster and IB have been funded by the excellence funds of the University of Heidelberg in the Frontier scheme. Though our parallel GPU code has not yet reached the perfection of standard NBODY6, we want to thank Sverre Aarseth for providing his codes freely and for teaching many generations of students how to use it and adapt it to new problems. This has helped and guided the authors in many respects.

REFERENCES

S. J. Aarseth. Star cluster simulations: The state of the art. *Celestial Mechanics and Dynamical Astronomy*, 73:127–137, 1999a.

S. J. Aarseth. From NBODY1 to NBODY6: The growth of an industry. *Publications of the Astronomical Society of the Pacific*, 111:1333–1346, 1999b.

S. J. Aarseth. *Gravitational N-Body Simulations*. Cambridge, UK: Cambridge University Press, 2003.

---

[7] http://iccs.lbl.gov.
[8] http://www.mao.kiev.ua/golowood/eng/.

A. Ahmad and L. Cohen. A numerical integration scheme for the N-body gravitational problem. *Journal of Computational Physics*, 12:389–402, 1973.

K. Akeley, H. Nguyen, and NVIDIA. *GPU Gems 3*. Addison-Wesley Professional, 2007.

J. Barnes and P. Hut. A hierarchical O(N log N) force-calculation algorithm. *Nature*, 324:446–449, 1986.

B. R. Barsdell, D. G. Barnes, and C. J. Fluke. Advanced architectures for astrophysical supercomputing. *ArXiv e-prints*, 2010.

R. G. Belleman, J. Bédorf, and S. F. Portegies Zwart. High performance direct gravitational N-body simulations on graphics processing units II: An implementation in CUDA. *New Astronomy*, 13:103–112, 2008.

P. Berczik, D. Merritt, and R. Spurzem. Long-term evolution of massive black hole binaries. II. Binary evolution in low-density galaxies. *The Astrophysical Journal*, 633:680–687, 2005.

P. Berczik, D. Merritt, R. Spurzem, et al. Efficient merger of binary supermassive black holes in nonaxisymmetric galaxies. *The Astrophysical Journal Letters*, 642:L21–L24, 2006.

P. Berczik, N. Nakasato, I. Berentzen, et al. Special, hardware accelerated, parallel SPH code for galaxy evolution. In *SPHERIC—Smoothed particle hydrodynamics european research interest community*, 2007.

P. Berczik, K. Nitadori, T. Hamada, et al. The parallel GPU N-body code φGPU. *in preparation*, 2011.

I. Berentzen, M. Preto, P. Berczik, et al. Binary black hole merger in galactic nuclei: Post-Newtonian simulations. *The Astrophysical Journal*, 695:455–468, 2009.

F. Chen, W. Ge, and J. Li. Molecular dynamics simulation of complex multiphase flows— Test on a GPU-based cluster with customized networking. *Science in China. Series B*, 38:1120–1128, 2008.

F. Chen, W. Ge, L. Guo, et al. Multi-scale HPC system for multi-scale discrete simulation. Development and application of a supercomputer with 1Petaflop/s peak performance in single precision. *Particuology*, 7:332–335, 2009.

H. M. P. Couchman, P. A. Thomas, and F. R. Pearce. Hydra: An adaptive-mesh implementation of P 3M-SPH. *The Astrophysical Journal*, 452:797, 1995.

Y. Cui, Y. Chen, and H. Mei. Improving performance of matrix multiplication and FFT on GPU. In *15th International Conference on Parallel and Distributed Systems*, 729:13, 2009.

W. Dehnen. A very fast and momentum-conserving tree code. *The Astrophysical Journal Letters*, 536:L39–L42, 2000.

W. Dehnen. A hierarchical O(N) force calculation algorithm. *Journal of Computational Physics*, 179:27–42, 2002.

E. N. Dorband, M. Hemsendorf, and D. Merritt. Systolic and hyper-systolic algorithms for the gravitational N-body problem, with an application to Brownian motion. *Journal of Computational Physics*, 185:484–511, 2003.

M. Fellhauer, P. Kroupa, H. Baumgardt, et al. SUPERBOX—An efficient code for collisionless galactic dynamics. *New Astronomy*, 5:305–326, 2000.

T. Fukushige, J. Makino, and A. Kawai. GRAPE-6A: A single-card GRAPE-6 for parallel PC-GRAPE cluster systems. *Publications of the Astronomical Society of Japan*, 57, 2005.

W. Ge and J. Li. Macao-scale pseudo-particle modeling for particle-fluid systems. *Chinese Science Bulletin*, 46:1503–1507, 2001.

W. Ge and J. Li. Macro-scale phenomena reproduced in microscopic systems-pseudo-particle modeling of fludization. *Chemical Engineering Science*, 58:1565–1585, 2003a.

W. Ge and J. Li. Simulation of particle-fluid systems with macro-scale pseudo-particle modeling. *Powder Technology*, 137:99–108, 2003b.

L. Greengard and V. Rokhlin. A fast algorithm for particle simulations. *Journal of Computational Physics*, 73:325–348, 1987.

A. Gualandris and D. Merritt. Ejection of supermassive black holes from galaxy cores. *The Astrophysical Journal*, 678:780–797, 2008.

T. Hamada and T. Iitaka. The Chamomile scheme: An optimized algorithm for N-body simulations on programmable graphics processing units. *ArXiv Astrophysics e-prints*, 2007.

S. Harfst, A. Gualandris, D. Merritt, et al. Performance analysis of direct N-body algorithms on special-purpose supercomputers. *New Astronomy*, 12:357–377, 2007.

R. W. Hockney and J. W. Eastwood. *Computer Simulation Using Particles*. Bristol, UK: Hilger, 1988.

C. Hou and W. Ge. GPU-accelerated molecular dynamics simulation of solid covalent crystals. *Molecular Simulation*, submitted, 2011.

W.-M.-W. Hwu. *GPU Computing Gems*. Morgan Kaufman Publ. Inc., 2011.

T. Ishiyama, T. Fukushige, and J. Makino. GreeM: Massively parallel TreePM code for large cosmological N-body simulations. *Publications of the Astronomical Society of Japan*, 61:1319, 2009.

A. Just, F. M. Khan, P. Berczik, et al. Dynamical friction of massive objects in galactic centres. *The Monthly Notices of the Royal Astronomical Society*, 411:653–674, 2011.

S. Komossa, V. Burwitz, G. Hasinger, et al. Discovery of a binary active galactic nucleus in the ultraluminous infrared galaxy NGC 6240 using Chandra. *The Astrophysical Journal Letters*, 582:L15–L19, 2003.

S. G. Lambrakos and J. P. Boris. Geometric properties of the monotonic lagrangian grid algorithm for near neighbor calculations. *Journal of Computational Physics*, 73:183, 1987.

K. J. Lee, N. Wex, M. Kramer, et al. Gravitational wave astronomy of single sources with a pulsar timing array. *ArXiv e-prints*, 2011.

B. Li, X. Li, Y. Zhang, et al. Lattice Boltzmann simulation on NVIDIA and AMD GPUs. *Chinese Science Bulletin*, 54:3178–3185, 2009.

D. Lynden-Bell and R. Wood. The gravo-thermal catastrophe in isothermal spheres and the onset of red-giant structure for stellar systems. *The Monthly Notices of the Royal Astronomical Society*, 138:495, 1968.

J. Makino. A modified Aarseth code for GRAPE and vector processors. *Proceedings of Astronomical Society of Japan*, 43:859–876, 1991.

J. Makino. A fast parallel treecode with GRAPE. *Publications of the Astronomical Society of Japan*, 56:521–531, 2004.

J. Makino and P. Hut. Performance analysis of direct N-body calculations. *The Astrophysical Journal Supplement Series*, 68:833–856, 1988.

J. Makino and S. J. Aarseth. On a Hermite integrator with Ahmad-Cohen scheme for gravitational many-body problems. *Publications of the Astronomical Society of Japan*, 44:141–151, 1992.

J. Makino, T. Fukushige, M. Koga, et al. GRAPE-6: Massively-parallel special-purpose computer for astrophysical particle simulations. *Publications of the Astronomical Society of Japan*, 55:1163–1187, 2003.

K. Nitadori and J. Makino. Sixth- and eighth-order Hermite integrator for N-body simulations. *New Astronomy*, 13:498–507, 2008.

S. Pasetto, E. K. Grebel, P. Berczik, et al. Orbital evolution of the Carina dwarf galaxy and self-consistent determination of star formation history. *Astronomy & Astrophysics*, 525:A99, 2011.

F. R. Pearce and H. M. P. Couchman. Hydra: A parallel adaptive grid code. *New Astronomy*, 2:411–427, 1997.

S. F. Portegies Zwart, R. G. Belleman, and P. M. Geldof. High-performance direct gravitational N-body simulations on graphics processing units. *New Astronomy*, 12:641–650, 2007.

M. Preto, I. Berentzen, P. Berczik, et al. Fast coalescence of massive black hole binaries from mergers of galactic nuclei: Implications for low-frequency gravitational-wave astrophysics. *ArXiv e-prints*, 2011.

H.-Y. Schive, Y.-C. Tsai, and T. Chiueh. GAMER: A graphic processing unit accelerated adaptive-mesh-refinement code for astrophysics. *Astrophysical Journal Supplement Series*, 186:457–484, 2010.

H.-Y. Schive, U.-H. Zhang, and T. Chiueh. Directionally unsplit hydrodynamic schemes with hybrid MPI/OpenMP/GPU parallelization in AMR. *The International Journal of High Performance Computing Applications*, accepted for publication.

V. Springel. The cosmological simulation code GADGET-2. *Monthly Notices of the Royal Astronomical Society*, 364:1105–1134, 2005.

R. Spurzem. Direct N-body simulations. *Journal of Computational and Applied Mathematics*, 109:407–432, 1999.

R. Spurzem, P. Berczik, G. Hensler, et al. Physical processes in star-gas systems. *Publications of the Astronomical Society of Australia*, 21:188–191, 2004.

R. Spurzem, P. Berczik, I. Berentzen, et al. From Newton to Einstein—N-body dynamics in galactic nuclei and SPH using new special hardware and astrogrid-D. *Journal of Physics Conference Series*, 780(1):012071, 2007.

R. Spurzem, I. Berentzen, P. Berczik, et al. Parallelization, special hardware and post-Newtonian dynamics in direct N-body simulations. In S. J. Aarseth, C. A. Tout, and R. A. Mardling, editors, *The Cambridge N-Body Lectures, volume 760 of Lecture Notes in Physics*, Berlin: Springer Verlag, 2008.

R. Spurzem, P. Berczik, G. Marcus, et al. Accelerating astrophysical particle simulations with programmable hardware (FPGA and GPU). *Computer Science—Research and Development (CSRD)*, 23:231–239, 2009.

R. Spurzem, P. Berczik, K. Nitadori, et al. Astrophysical particle simulations with custom GPU clusters. In *10th IEEE International Conference on Computer and Information Technology*, pp. 1189, 2010.

R. Spurzem, P. Berczik, T. Hamada, et al. Astrophysical particle simulations with large custom GPU clusters on three continents. In *International Supercomputing Conference ISC 2011, Computer Science—Research and Development (CSRD)*, accepted for publication, 2011.

A. C. Thompson, C. J. Fluke, D. G. Barnes, et al. Teraflop per second gravitational lensing ray-shooting using graphics processing units. *New Astronomy*, 15:16–23, 2010.

P. Wang and T. Abel. Magnetohydrodynamic simulations of disk galaxy formation: The magnetization of the cold and warm medium. *The Astrophysical Journal*, 696:96–109, 2009.

P. Wang, T. Abel, and R. Kaehler. Adaptive mesh fluid simulations on GPU. *New Astronomy*, 15:581–589, 2010a.

X. Wang, W. Ge, X. He, et al. Development and application of a HPC system for multi-scale discrete simulation—Mole-8.5. In *International Supercomputing Conference ISC10*, 2010b.

H.-C. Wong, U.-H. Wong, X. Feng, et al. Efficient magnetohydrodynamic simulations on graphics processing units with CUDA. *ArXiv e-prints*, 2009.

Q. Xiong, et al. Large-scale DNS of gas-solid flow on Mole-8.5. *Chemical Engineering Science*, submitted, 2010.

G. Xu. A new parallel n-body gravity solver: TPM. *Astrophysical Journal Supplement Series*, 98:355, 1995.

J. Xu, Y. Ren, X. Yu, et al. Molecular dynamics simulation of macromolecules using graphics processing unit. *Molecular Simulation*, submitted, 2009.

J. Xu, H. Qi, X. Fang, et al. Quasi-realtime simulation of rotating drum using discrete element method with parallel GPU computing. *Particulogy*, in press, 2010a.

J. Xu, X. Wang, X. He, et al. Application of the Mole-8.5 supercomputer—Probing the whole influenza virion at the atomic level. *Chinese Science Bulletin*, in press, 2010b.

R. Yokota and L. Barba. Treecode and fast multipole method for N-body simulation with CUDA. *ArXiv e-prints*, 2010.

R. Yokota, J. P. Bardhan, M. G. Knepley, et al. Biomolecular electrostatics using a fast multipole BEM on up to 512 GPUs and a billion unknowns. *ArXiv e-prints*, 2010.

K. Yoshikawa and T. Fukushige. PPPM and TreePM methods on GRAPE systems for cosmological N-body simulations. *Publications of the Astronomical Society of Japan*, 57:849–860, 2005.

Chapter *4*

An Overview of the SimWorld Agent-Based Grid Experimentation System

Matthias Scheutz

Department of Computer Science, Tufts University, Medford, MA, USA

Jack J. Harris

Human Robot Interaction Laboratory, Indiana University, Bloomington, IN, USA

4.1 INTRODUCTION

Computational modeling is becoming increasingly important, even in fields that have not traditionally used computational models (e.g., archaeology or anthropology). Researchers in both the natural and social sciences employ computer simulations to elucidate the time course of physical and nonphysical processes, or to explore the dynamics among different interacting entities, in an effort to discover new relationships that might lead to generalizable laws or to verify hypothesized principles as part of the empirical discovery loop (Peschl and Scheutz, 2001). However, there are two main obstacles to making effective use of today's (and likely also tomorrow's) computing environments. First, navigating the complexity associated with running large-scale

Large-Scale Computing, First Edition. Edited by Werner Dubitzky, Krzysztof Kurowski, Bernhard Schott.
© 2012 John Wiley & Sons, Inc. Published 2012 by John Wiley & Sons, Inc.

computational simulations requires detailed knowledge about the available high-performance computing (HPC) environments. Such prerequisite knowledge includes how to set up a simulation on the host computers (possibly including compilation on the target platform with installation of all the required libraries), how to schedule sets of simulations through the batch system, how to retrieve the resultant data, and how to troubleshoot if simulations do not finish (because they were terminated by the cluster's batch system for taking up more than the allocated CPU time, memory, or storage allotment). The second obstacle is the management of increasingly large data sets that are the result of explorations of larger and larger model parameter spaces, both in terms of the dimensionality and sampling density of the space. This includes the preprocessing of data to facilitate statistical analysis and data mining, and the visualization of interesting relationships among data sets. Either one of these obstacles is usually prohibitive for nonexperts and will ultimately prevent modelers from using HPC resources to run large-scale simulations.

While there are certainly other challenges involved in making computational modeling more accessible to nonprogrammers (e.g., better modeling environments and tools for developing computational models in the first place), in this chapter, we will focus on the above-mentioned two challenges related to the HPC environment and the subsequent data analysis and visualization phase. Our goal is to make the computational modeling process in HPC environments as easy and intuitive as possible for modelers. This includes providing a computational framework that can automatically schedule, parallelize, distribute, and run simulations using different strategies for the exploration of large parameter spaces. Moreover, it includes tools for automatically collecting data, organizing data in databases (that enable efficient data mining and statistical analyses), and visualizing data in effective, easily specifiable ways. Ultimately, we would like to have a framework that supports the *entire computational modeling process* (Peschl and Scheutz, 2001): from developing the first model to testing and running it, to collecting, analyzing, and visualizing data, to comparing data to empirical findings, revising the model, testing it, and so forth. Furthermore, this infrastructure should attempt to minimize model run times to speed up this process, for example, by automatically parallelizing the model (as we do not want to require modelers to be able to implement parallel code that can be executed on a cluster). Moreover, it would be desirable if the infrastructure could automatically handle vastly heterogeneous computing infrastructures (e.g., from dedicated homogeneous HPC environments to heterogeneous ad hoc clusters with different operating systems). It would also be desirable for the infrastructure to dynamically adjust to changing computing environments since computational resources and resource availability can vary greatly over time and across research settings, and typically, very specific knowledge is required to schedule and run processes in each environment.

To this end, we present SimWorld Agent-Based Grid Experimentation System (SWAGES). SWAGES has been under development for over a decade in our lab. SWAGES is used extensively for various kinds of agent-based modeling. In particular, SWAGES was codeveloped with SimWorld[1] (Scheutz, 2001), an agent-based modeling environment built on top of the Birmingham SimAgent agent toolkit (Sloman and Logan, 1999). SimWorld is a generic simulation environment for spatial agent-based models that provides both interactive and batch mode execution and permits the definition of agent-based models in several programming languages. It has been used extensively for simulations of artificial life scenarios (e.g., Scheutz and Schermerhorn, 2008), evolutionary investigations (e.g., Scheutz and Schermerhorn, 2005), social simulations (e.g., Scheutz and Schermerhorn, 2004), swarm-based simulations (e.g., Scheutz et al., 2005), and individual-based biological models (e.g., Scheutz et al., 2010). It has also been used in education for teaching model exploration and model development (e.g., Scheutz, 2008). SimWorld was the first simulation environment to support the automatic parallelization algorithms specified by and implemented in SWAGES (Scheutz and Schermerhorn, 2006). While the reimplementation of SimWorld in Java is still under development, several additional asynchronous scheduling algorithms have already been included. Evaluation of these asynchronous scheduling policies has demonstrated performance gains compared to the typical cycle-based scheduling policy used in the previous version of SimWorld and most other discrete-event simulators (Scheutz and Harris, 2010).

SWAGES can be used to explore large parameter spaces of various models of cognition, both connectionist models and classical cognitive architectures. On the connectionist side, our neural network simulator NNSIM has been used to explore parameter spaces of several neural networks, including neural networks for spatial attention (Scheutz and Gibson, 2006) and ideomotor compatibility (Boyer et al., 2009). On the cognitive side, a special-purpose Lisp-based ACT-R "wrapper component" was developed to run the ACT-R cognitive architecture (Anderson et al., 2004) and to pass various model parameters between the cognitive model and SWAGES, thereby enabling automatic explorations of large parameter spaces (e.g., using various ACT-R models in the "psychomotor vigilance task"; Gluck et al., 2007).

SWAGES started out as a set of shell scripts for running agent-based artificial life models on remote hosts in the late 1990s and has since evolved into a robust, advanced modeling framework that meets all of the above-mentioned requirements. The following list highlights some of its features:

1. SWAGES runs on any platform that supports Java and can be automatically distributed over multiple hosts to facilitate high throughput for large numbers of parallel simulations.
2. It use a heterogeneous computing environment including any mixture of dedicated compute clusters or stand-alone hosts with no preinstalled software required on any host (only secure shell access is needed).

---

[1] SimWorld was the first simulation environment supported by SWAGES, hence the "SimWorld" prefix in SWAGES, even though it now works with many simulation environments.

3. It works with any simulation environment (including closed-source simulations) that can be minimally parameterized (e.g., through command-line arguments or a special-purpose socket-based protocol).

4. It can automatically parallelize simulations based on available computational resources for simulation environments that support parallelization, including synchronous and asynchronous scheduling algorithms, to maximize throughput using a dynamic pool of hosts.

5. It provides a simple intuitive Web-based user interface for specifying, scheduling, and running large-scale model parameter spaces (including different supplied exploration algorithms as well as user-defined strategies that can automatically schedule additional simulations based on simulation outcomes).

6. It enables automatic data retrieval and population of databases for efficient data mining.

7. It facilitates automatic statistical analyses, which include both model fitting based on fitness criteria (error thresholds, types of models, etc.) and model discovery (based on constraints on model classes).

8. It enables automatic visualization of different data sets obtained from large-scale simulations (and for linking in other visualization environments).

9. It contains mechanisms for ensuring that simulations will eventually finish (despite crashes, interrupted simulations, or lack of available hosts), including checkpointing mechanisms if supported by the simulation model.

SWAGES is described in detail in the following sections, starting with an overview of the system architecture. Next, the new implementation of SWAGES in the distributed Agent Development Environment (ADE) (Scheutz, 2006) is discussed and some of the advantages of distributing its architecture are highlighted. Then, we describe an application of SWAGES in the context of a biological agent-based model to highlight how SWAGES addresses the challenges of exploring large model parameter spaces in HPC environments. Finally, we compare SWAGES to related large-scale simulation frameworks and summarize its new features.

4.2 SYSTEM ARCHITECTURE

The initial design goal for SWAGES is to simplify the modeling process by allowing modelers to define model parameter spaces they want to explore and to submit them to SWAGES for execution. SWAGES then automatically schedules all simulations, runs them, collects the resultant data, and transfers them to a specified place in the file system. It also performs simple statistical analyses and displays them through a Web-based user interface. Figure 4.1

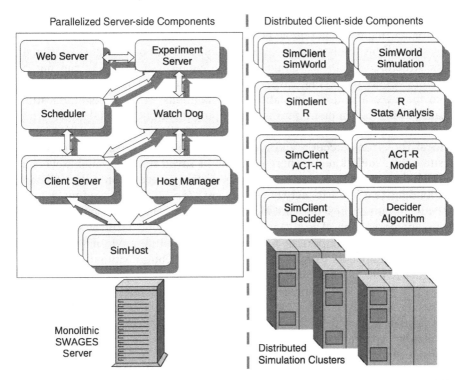

Parallelized Server-side Components | Distributed Client-side Components

Figure 4.1 *Overview of the SWAGES architecture in 2009 (all server-side components within the solid square on the left had to run on the same host computer).*

shows an overview of the SWAGES architecture divided into "server-side" and "client-side" components as described by Scheutz et al. (2006) and Scheutz (2008).

The server-side components are responsible for scheduling, distributing, starting, and monitoring the execution of distributed simulation clients and possibly restarting failed clients to ensure that simulations will eventually finish. Specifically, users can submit experiments, check their status, perform simple statistics, and view experiment results using any standard Web browser through a Web-based user interface provided by the *Web server*. The Web server forwards all submission requests to the *experiment server*, which creates "experiment sets" based on user specifications that contain all necessary model parameters (e.g., the set of initial conditions across different experiments) as well as various SWAGES system parameters. System parameters can include priorities of experiments and scheduling parameters, levels of supervision and recovery parameters, formats of data collection and location for data storage, statistical analysis calculations to be used on the results and conversions of the output formats, and modes of user notification of simulation progress. The *scheduler* is responsible for taking experiments from several priority-based queues (to which new experiments are submitted by the experiment server) and for starting them on remote hosts. The experiment scheduler will schedule new simulations when hosts become available and will only create experiment data structures for these simulations on demand (so as not to run out of memory when processing large-scale experiment sets). The *client server* then manages a remote simulation and maintains an open communication channel with the simulation instance, keeping track of the simulation's progress, state, update, and degree of parallelization. This ongoing monitoring is critical for error detection and recovery so the client server can restart or resume the simulation elsewhere based on its saved state, if available, when a simulation crashes (e.g., due to operating system problems on its host), is not responding (e.g., due to network problems), or cannot be continued (e.g., because its current host does not meet user-defined criteria for running simulations anymore). The *Watch Dog* implements a second level of supervision that is particularly important for dynamic computing environments where hosts can "disappear" from the pool of usable machines without notification. It regularly checks all simulation clients for progress, terminates clients that are stuck or are not responsive, and reschedules simulations either from scratch or from saved states. The server-side representation of remote simulation hosts is achieved by the *SimHost* component, which keeps track of any simulations running on the host and is also responsible for monitoring the host's availability based on user-defined criteria (e.g., the remaining CPU time on a cluster host or whether a console user is logged on). The *host manager* keeps track of all available simulation hosts by managing a dynamic pool of available resources. It can automatically request new hosts in HPC environments by submitting requests on demand through the cluster's batch queuing system.

On the client side, special *SimClients* representing particular simulation environments communicate updates about the simulations to the (server-side) client servers. SimClients are responsible for saving the state of simulations (if supported by the simulation) and for routinely checking that the simulation is still allowed to run on its current host according to user-defined criteria. Generic SimClients are available to interface with simulations written in various programming languages (Java, Pop11, Scheme, Lisp, and R), and customized SimClients are available for SimWorld (Scheutz, 2001) and ACT-R (Anderson et al., 2004). Furthermore, SWAGES also supports various special-purpose reusable clients (e.g., a generic *gradient search client*[2]).

Recently, however, it has become clear that to truly scale up to the requirements of tomorrow's modelers, several additional steps in the design and implementation of SWAGES need to be taken. First, the server-side part of

---

[2] This client is capable of wrapping a generic simulation process and supports batching a group of parameters to run on the client. This client has a twofold benefit: (1) It minimizes the amount of independent server connections, which can incur an overhead for very fast simulations and (2) aids distributed Monte Carlo search experiments.

SWAGES, while parallelized, was monolithic (i.e., all server-side components had to run within the same Java virtual machine on the same host). And while multicore CPUs are partly able to alleviate the *processing bottleneck*, they cannot help the *networking bottleneck* created by a monolithic grid engine that communicates with thousands (if not tens of thousands) of simulations simultaneously. The solution, therefore, is to distribute and duplicate (some of) the parallelized server-side components over multiple hosts. Second, for SWAGES to effectively handle large data sets (of hundreds of gigabytes and beyond), simply storing data as text files in the file system is not practical as data queries searching sequentially through these large files would take too much time. Rather, a database interface is required that allows automatic commits of data (as results are produced by SimClients) into a database (which potentially can be distributed itself and can facilitate efficient data querying and data mining). Third, to aid the modeler in the exploration of large parameter spaces, automatic statistical analyses on the returned intermediate data sets are required in order to determine whether a particular region of the parameter spaces should be further explored. Hence, additional mechanisms for automatic data analysis, model fitting and model discovery were developed and integrated. Fourth, the previously integrated simple visualization mechanisms were too limited to do justice to the complexities of large data sets. Therefore, a new visualizer component has been added to provide better, more effective automatic generation of data visualizations based on model space parameters and experimental results. Finally, to reduce the complexities of data exchanges between different components (from the simulation environment to the database, statistical analysis, and visualization) and to enable the easy integration of external components (e.g., simulation environments, statistical analysis tools, and visualizers), a new open, experiment definition and data exchange format based on the extensible markup language (XML) was specified and implemented.

To address the required modification of SWAGES (for scaling up to large numbers of simultaneous simulations), all SWAGES components were reimplemented in the distributed ADE (e.g., Scheutz, 2006). ADE provides many features that are of critical importance for the distributed version of the SWAGES server-side components: automatic load balancing and host management, component supervision, error detection and restart, and various other mechanisms for *autonomic computing*. Moreover, ADE's tight security services provide fine-grained, method-level authentication, which is integtral in an open distributed computing environment. ADE also provides distributed graphical interfaces that allow location-independent system configuration and monitoring. To leverage these features, SWAGES components were converted into ADE components and new features related to large-scale computational simulation experimentation were added. At a high level, SWAGES is now composed of the following components: the *manager*, the *engine*, the *analyzer*, and the *visualizer* (the extended server-side architecture of SWAGES as implemented in ADE is shown in Fig. 4.2).

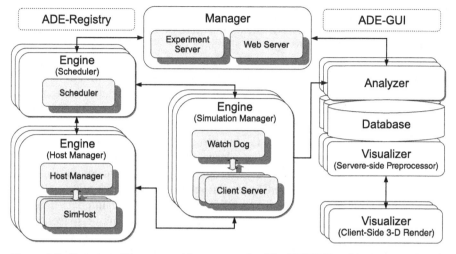

Figure 4.2   *Overview of the server-side components of the SWAGES architecture as currently implemented in ADE. ADE components can be freely distributed to any host, thus enabling the distribution of SWAGES server-side components. Except for the manager, multiple copies of each component are now possible within a SWAGES instance, thus allowing SWAGES to scale with the number of simulations it is serving (see text for detail). Large, rounded boxes depict new ADE components; small, gray rounded boxes depict old SWAGES components; the two dashed, rounded boxes show ADE infrastructure components.*

The manager is responsible for receiving experiment specifications and for potentially ending a simulation based on specified termination criteria. It initializes the engine and the analyzer together so they can communicate directly with each other in order to process simulation results.

The engine receives the specification of a simulation, including simulation execution details and the parameter space for evaluation from the manager, and is responsible for distributing the simulation environment across a collection of networked nodes and for consolidating the results. The engine is considered a reliable service and will ensure that results are returned from a simulation by handling events such as checkpointing and restarting of simulations.

As results are collected by the engine, they are forwarded to the analyzer. The analyzer first stores the data in an indexed database to aid data analysis and future interactive exploration of the results. The analyzer then performs a battery of statistical analyses on the data. Topographical function forms (which can potentially explain the relationship of parameters in an experiment) are produced for the specific parameter space and are evaluated in real time using distributed statistical processing nodes as the data returns from the simulations. The analyzer also provides a conduit for carrying out post hoc data evaluation by data mining over the processed results (e.g., by supplying an interface for querying the simulation results closest to a particular parameter combination). This type of post hoc data mining of a parameter space lets modelers identify optimal models during model validation and evaluation.

By complementing the analyzer, the visualizer can provide a way for quickly identifying interesting parameter relationships, understanding the breadth of a simulation's performance or even identifying the problematic performance of a simulation through interactive three-dimensional renderings of the processed data. The visualizer can plot individual data points, line plots, lego plots, and even surface plots in order to help explain the properties of the performance space.

4.3 SYSTEM IMPLEMENTATION

The ADE middleware for designing distributed autonomous agent systems was chosen as the implementation platform for the extended version of SWAGES for several reasons. First, a number of other higher-level architectures have already been successfully developed within the ADE framework including DIARC, an embodied agent framework for robot software development (Scheutz et al., 2007) and ADE-Unreal, a virtual world simulator. Thus, ADE provides a suitable infrastructure that allows functionality associated with SWAGES to be distributed across a set of computers. ADE is also implemented in Java, which aids easy code integration of existing SWAGES components. Furthermore, this distributed infrastructure permits the introduction of new processor- and storage-intensive functionality associated with large-scale computational experimentation.

To facilitate the integration of the separate functional components, each existing and new SWAGES service is implemented or "wrapped" as an ADE component. This is possible because ADE provides a class hierarchy for extending ADE services, which gives new services the ability to intercommunicate with one another as well as utilize core ADE features (such as server security, a server recovery mechanism, service logging, and monitoring). Access to ADE services is readily accessible using this wrapping mechanism; however, non-Java-based services can also intercommunicate with ADE servers using Web services in the form of XML-RPC[3] or Representational State Transfer (Fielding and Taylor, 2002). New ADE services extending *ADEWebServicesServer* permit the exposure of a subset of remote methods based on user credentials and remote IP.

ADE enables secure message transport to SWAGES components via a remote method calling feature that utilizes Java's Remote Method Invocation, a set of application programming interfaces that allow developers to build distributed applications. ADE also provides a metalevel naming scheme to

---

[3] http://xmlrpc.com/spec.

allow an abstraction over these interfaces capable of service migration. The remote method calling feature of ADE is protected using a centralized authentication and access system. This is the predominant mechanism by which ADE services communicate.

ADE also provides a method for creating distributed graphical user interfaces (GUIs) using ADE-GUI, an extensible framework that abstracts away many of the issues of distributed GUI development. This abstraction and encapsulation of functionality thereby aids the easy creation and integration of new GUI components. Many components within the system implement GUIs that can be viewed through ADE-GUI components; however, some servers provide an additional Web-based user interface for control.

4.3.1 Key Components

4.3.1.1 *Messaging* The SWAGES system communicates using our new Simulation Specification Markup Language (SSML). This simple XML specification allows all components to utilize the same data structures for defining simulation experiments. Such a shared representation is useful because many components need access to the same type of information (e.g., the analyzer and the visualizer both need access to schemas describing the results data). There are primarily three types of top-level SSML structures: execution details, parameter space definition, and the results definition schema. By using these basic structure, SWAGES components can communicate with one another (using the standard message passing mechanisms defined within ADE).

4.3.1.2 *Manager* Centralized access to the collection of SWAGES components is provided through the manager. The manager is responsible for receiving all of the components' specifications used to process, analyze, and evaluate the experiments. This role is similar to that of the experiment server in the previous version of SWAGES. The manager receives the three basic SSML structures from a user and passes them along to other SWAGES components. The manager also relays to the engine simulation execution details along with the parameter space information, which together define the scope of the simulations to evaluate. User interfaces to the manager are provided via the command line and other external add-on ADE components in the ADE network. The manager also exposes programming methods using Web services to make the direct submission from external user applications possible.

4.3.1.3 *Engine* The engine is essentially a distributed form of the server-side SWAGES components with several extensions to let it communicate externally with other ADE services. The engine is composed of three distinct ADE components that allow server-side components of SWAGES to run on multiple hosts. These three new components of the engine communicate with each other in order to carry out the task of distributing and managing the execution of simulations.

4.3.2 Novel Features in SWAGES

Next, we briefly summarize the advantages of distributing the SWAGES architecture and its extended features that allow it to interoperate with the other components.

4.3.2.1 Distributed SWAGES
The distribution of the subcomponents of the engine across a set of hosts provides many performance advantages. As the number of executing SimClients grows, so, too, does the amount of communication back to the SWAGES server. Large amounts of concurrent communication can potentially render a client server as a bottleneck for a parallel simulation run. Processing these messages can be both a processor-intensive operation as well as a bandwidth bottleneck. Distributing the management in a hierarchical manner produces a completely scalable system (e.g., a new client server can be started on a new host as soon as existing ones reach their capacity limits based on experiment submissions and available simulation hosts).

4.3.2.2 Postprocessing
The system provides additional components to process the results returned by the remote simulations. These postprocessing tools provide services including data warehousing, statistical analysis, and visualization features. Some of the services rely on system-unique resources such as database installation or even graphic card requirements. However, as is the case for all ADE components, these services do not need to run on the same host as the engine or the manager but can run on any host in the ADE environment that meet the requirement of the service.

4.3.2.3 Analyzer/Database
It is not uncommon for a large-scale distributed simulation experiment to evaluate billions of parameter combinations and to produce gigabytes (if not terabytes) of resulting data. As Charlot et al. (2007) point out, "These data need to be managed, shared and analysed using varied computational methodologies, such as data mining and database management systems." SWAGES can perform such tasks through the analyzer, which provides the storage and efficient management capabilities to facilitate automated results analysis as well as real-time interactive data mining.

4.3.2.4 Database
Loading all simulation results into the main memory for analysis quickly becomes infeasible with large-scale simulations producing terabytes of data; even searching for a desired result in a massive data set (e.g., over 1 billion simulation results) can be extremely time-consuming if standard sequential file access is used to scan a file system. The analyzer, therefore, utilizes a MySQL[4] database to house the massive number of results produced by SWAGES. The tables are indexed using "B+ trees" (Comer, 1979), which minimize the number of file system accesses required to search for a result

---

[4] http://mysql.com.

while also maximizing the usage of the main memory of the system. Users can manage and manually analyze a stored data set by accessing it via a Web-based user interface.

4.3.2.5 Analysis Processes   With a robust data storage and retrieval backend in place, the analyzer can provide automated analysis of large amounts of data. The processes for doing this analysis involve creating and populating a model repository (database), learning topographical functions that best fit the data (automated analysis), and executing predictions and reverse lookups from the data store (interactive analysis).

The process begins when information about the parameter space is received from the manager and a model repository is created to house the data. A model repository is defined in terms of the variables of the parameter space for the experiment. Using additional information provided by the manager, the analyzer then subscribes to the engine and requests all results produced for a given experiment. The received data is parsed and stored into the respective model repository for that experiment. The analyzer then begins to produce a set of potential "hypotheses" about the functional relationship between the parameters of the space.

As results are uploaded from the remote simulations, automated analysis of the data begins. The goal of the analysis is to find a function that best describes the topographical form of the data based on the set of function schemas originally generated in the previous process. Multivariate regression analysis, which uses the database directly, is used to evaluate each candidate function. Earlier versions of the analyzer utilized multiple subprocesses running R[5] and MATLAB[6] for statistical analysis. Though effective, memory constraints for very large data sets produced run times an order of magnitude slower than the database statistical package written for the current analyzer. Thus, relying on the database directly versus using external statistics packages has proven to be an efficient way of doing automatic data analysis for large data sets.

As new results flow into the system and the data set grows, the candidate topographical function forms are continuously reevaluated based on the goodness of fit to the current data set. The selection of which functions to evaluate is determined by a greedy algorithm that identifies those function forms that best fit the data set prior to the addition of the new results. The result of this process is that at any moment, the analyzer has a set of functions that describe the relationship of the parameters for the experiment.

4.3.2.6 Visualizer   The visualizer provides an interactive three-dimensional graphical interface, using the FreeHEP Java3D[7] library extensions (produced

---

[5] http://cran.r-project.org.
[6] http://www.mathworks.com/products/matlab/.
[7] http://java.freehep.org.

by the high-energy physics modeling community), for plotting model parameters against one another to aid the evaluation process. However, visualization, like data analysis, is complicated by the problem of handling vast amounts of data; therefore, the visualizer provides preprocessing mechanisms for data to minimize the amount of data that needs to be sent to the user's three-dimensional rendering system. Hence, the visualizer consists essentially of two components: the client-side rendering application and the database plus the preprocessing server-side scripts. This architecture makes the visualization of massive data sets possible and efficient by exploiting the speed of the highly indexed database representation and by minimizing the data that need to be transferred to the client.

4.4 A SWAGES CASE STUDY

In this section, we present an interdisciplinary research project in biology and computer science that demonstrates the use and application of SWAGES. We first summarize the research questions addressed in the particular project and briefly introduce the agent-based simulation environment used to explore these questions. Then, we discuss the role SWAGES played in conducting the simulation experiments and the subsequent data analysis, showing how large-scale simulation experiments can be defined and executed in a grid environment and ultimately, how the results returned from the parallel simulations can be stored, analyzed, and visualized to provide feedback to the modeler.

4.4.1 Research Questions and Simulation Model

The main research question we explored in the project comes from mate choice in biology: How do females pick their mates? Specifically, how do female tree frogs decide which male tree frog to approach in the swamp at night when the only information about males available to females are the acoustic properties of the males' mating calls? Prior work in theoretical and ecological biology proposed two main competing strategies for females: "Pick the closest out of N males" or "pick the closest above some threshold θ." To test the success of and the trade-offs among these two strategies, we developed a social agent-based model that explicitly models male and female frogs as "agents" located in a spatially extended, two dimensional environment. The model included both environmental parameters (e.g., the number and distributions of male and female frogs) and individual parameters (e.g., the call quality of males or the mating strategy employed by females). The goal was to systematically vary environmental and individual parameters to gain an understanding of the various dependencies and trade-offs among the different dimensions (e.g., how mating success depended on the distribution of males, how time to mating depended on the number of competing females, and how the male–female ratio influenced the performance of the different strategies).

Details about the biological questions and the computational model have been reported by Scheutz et al. (2010).

4.4.2   The Simulation Environment

The social female choice model was implemented as a discrete-event agent-based simulation in Java. After initial testing of the simulation in a graphical environment to verify that all agents individually behaved as expected, the simulation environment was connected to SWAGES to permit the automatic scheduling and running of the millions of simulations required for the parameter space we had set out to explore. The integration of the simulation with SWAGES was accomplished by extending a SWAGES client-side component, the *JavaClient*, and by overriding some of its methods. The primary method that had to be overridden was the entry procedure, which receives a list of start-up parameters used by the simulation to govern its operations (other methods can also be overridden if advanced features such as checkpointing or SWAGES-level control over the event-scheduler algorithm are desired). The main functionalities provided by the JavaClient are the systematic representation of agent types and the parallelizable discrete-event scheduler (details of an integration that takes advantage of the advanced intrasimulation discrete-event scheduling algorithms are described by Scheutz and Harris, 2010). These predefined algorithms allow for quick model development and verification. Moreover, the social female choice simulation environment used mechanisms provided by the JavaClient for handling initial conditions and for specifying event details (e.g., mating events) for data recording during simulation runs. All data are recorded in systematically named files, which can then be collected by SWAGES for subsequent data analysis.

4.4.3   Simulation Runs in SWAGES

The data presented in this case study were obtained by operating SWAGES within a stand-alone HPC configuration where both the server and processing nodes all operated on the BigRed supercomputer.[8] Other phases of the research project used different HPC configurations (e.g., a heterogeneous set of processing nodes consisting of local lab machines and compute nodes on a shared resource grid). Regardless of the particular composition of the computing environment, the simulation environment, along with resource files defining the parameter space, the simulation's execution details, and the resulting output files' data format, needs to be submitted to the SWAGES server. SWAGES then initializes simulations on compute nodes as they become available, passing start-up parameters to simulation instances and collecting data from simulation output files for subsequent insertion into the SWAGES database.

---

[8] For details on the BigRed system, see http://kb.iu.edu/aueo.html.

For the particular experiment set performed on BigRed, three different mating strategies were evaluated: a *minimum threshold* strategy (where the closest mate exceeding some fitness criteria is selected), a *best of* N strategy (where the mate with the highest fitness metric is chosen out of the N closest candidate mates), and a random mating strategy (used as a baseline). Each strategy was tested across the same set of initial starting conditions which varied different environment factors and the characteristics of the frog population. For example, for the random condition to be evaluated across the range of initial conditions, 240,000 simulations were run. More simulations were run for the minimum threshold and best of N because both strategies included additional strategy parameters (fitness threshold level and size of N, respectively). For the best of N strategy, five different values of N were evaluated. This resulted in 1,200,000 simulation runs ($5 \times 240,000$). For the minimum threshold strategy, 10 different values for the fitness threshold were tested, resulting in 2,400,000 additional simulation runs ($10 \times 240,000$).

In order to execute such a large number of simulations quickly, a large number of compute nodes were required. However, at the time of submission, the queue on the cluster environment was long for simulations with a large number of compute node requests. Rather than attempting to reserve a large number of nodes that would have caused the requests to sit in the queue for weeks, we decided to let SWAGES request compute nodes dynamically and individually (up to a maximum number of 50 node requests at a time each for 4-hour increments). When nodes became available, SWAGES then scheduled and executed simulations until that node's allotted time expired.

4.4.4 Data Management and Visualization

As a result of employing this dynamic resource request strategy, the entire set of 3,840,000 simulations finished in a little over 2 weeks from the time they were submitted (roughly 17 days). The simulations generated over 11 gigabytes of summary data (with precomputed statistics) and over 10 terabytes of detailed results in over 35 billion files. Clearly, without tools like SWAGES, the number of simulations and the size of the resultant data would be prohibitive for a model exploration of this magnitude. For example, for large data sets like the ones produced in this study, standard file systems are not a practical method for data management when searching for patterns inside the data files requires a slow scan of every file. This is where the automatic insertions of results in a database has proven to be of great use for quickly querying the data set. Earlier versions of SWAGES did not utilize a database. Instead, custom scripts were generated to search the returned result files for patterns of interest. To appreciate the potential speedup provided by the automatic database creation and usage, consider the time it would take to search through every result file to find the set of parameters that produced the maximum number of mated tree frogs: The process of using result files involved searching each file, which required N file accesses, where N is the total number

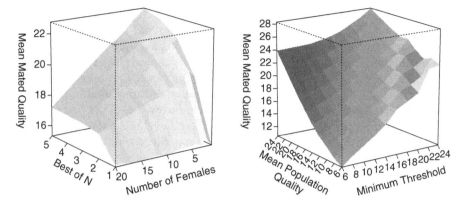

Figure 4.3 *Sample graphs of results generated based on a systematic exploration of the parameter space for two strategies. The left graph shows the dependence of the quality of the mated male frogs on the number of females in the simulation and different parameters for the best of N female mating strategy. The right graph shows the dependence of the quality of the mated male frogs on their average call qualities and the value for the minimum threshold in the female mating strategy.*

of parameter combinations explored. In the case where the data are stored in an indexed database, the desired value can quickly be found without the need to examine every simulation result.

Once the data are preprocessed, visualizations can be produced. In previous versions of SWAGES, these visualizations had to be generated manually; in the current version, visualizations can also be produced automatically. Figure 4.3 shows examples of the simulation results from the frog study that visualize different effects of mating strategies. These types of graphs are immediately meaningful for the modeler and have been used both for verifying theoretical predications as well as developing new, refined computational models.

4.5 DISCUSSION

Computational frameworks and infrastructures that are intended to support large-scale explorations of model parameter spaces face significant challenges as the numbers of usable CPUs and hosts in clusters and grid environments rapidly grow and data sets increase in size by several orders of magnitude. The first question that arises is how an infrastructure will "scale up" in light of this enormous growth in both resources and resource demands. A second related question is how the large amounts of data should be managed (i.e., stored, accessed, analyzed, mined, and visualized). And third, as infrastructures become larger and more complex, a technical question arises as to whether and to what extent these infrastructures are able to autonomously handle various types of errors (component crashes, network problems, etc.) that are inevitable in complex computing systems. The goal of the reimplementation of SWAGES in the ADE middleware is to answer all three questions and to address the associated challenges.

4.5.1 Automatic Parallelization of Agent-Based Models

One feature that makes SWAGES unique among experimentation frameworks is its ability to automatically parallelize and distribute agent-based models. This feature is particularly important for modelers who would like to quickly run a particular model, visualize the results, possibly change a few parameters, and run it again. This type of *interactive model exploration* is usually only possible with models that run over a short period of time on a single computer, but not with models that have run times of tens of minutes or hours (when minutes might already be too long). Since there is a natural limit to how fast a simulation can run on a single computer, the only way to allow quasi-real-time interactions is to parallelize the model. However, parallelization usually requires advanced programming skills *and* support by the modeling environment (neither of which is usually available). SWAGES addresses this problem by removing the burden of parallelization from the modeler. Specifically, the need for a modeler to manually execute model simulation runs is eliminated by SWAGES with its ability to automatically utilize all computational resources available for parallelization. This is possible for simulation environments that implement an "event horizon," a particular notion of influence a simulated entity can have on other simulated entities within the model simulation (Steinman, 1994; Scheutz and Schermerhorn, 2006). While the event horizon is currently only implemented for (metric) spatial agent-based models, it can be generalized to nonspatial models. For example, it could be used for nonspatial models that are based on interaction graphs that connect every entity in a simulation to those entities with which it can interact.

In addition to parallelization, SWAGES also provides mechanisms to support *asynchronous updating and scheduling within the simulation environment*. Parallelization allows simulation instances to run asynchronously on different hosts until one simulation instance needs data from an entity that is updated in another simulation instance. In asynchronous scheduling, simulation entities are not updated cycle by cycle but are scheduled (to the extent possible) based on which entity in one simulation instance will be required to provide the information needed by another simulation instance in the future. This anticipatory way of updating entities within each simulation instance in the distributed simulation system can lead to significant overall run-time speedups (Scheutz and Harris, 2010).

In general, the automatic parallelization of simulation models enables fast "model discovery loops" (especially with complex models) while alleviating the modeler from having to parallelize model code manually. Furthermore, parallelized models also facilitate distributed real-time simulations (e.g., of mixed live–virtual–constructive environments) that would otherwise not be able to run in real time.

4.5.2 Integrated Data Management

A large amount of data can be produced by even simple agent-based models. For example, models created by Scheutz et al. (2010) produced hundreds of gigabytes even though each model run finished within only a fraction of a second. In such cases, it is not possible to store simulation data distributed over millions of files in the regular file system (e.g., we have seen simulations produce more result files than a directory could hold given the file system restrictions of the operating system). Rather, data need to be automatically inserted in a database in a way that aids efficient data access. SWAGES automatically creates a database based on the initial experiment setup and inserts simulation results in an organized way into tables that can subsequently be combined using Structured Query Language queries. Moreover, since the database can efficiently access large data sets, it is used to preprocess the data for real-time interactive visualizations. The immediate availability of visualizations (if only on sparse data) lets modelers anticipate results and possibly correct simulation setups early on (e.g., if the explored region of the space is not interesting or if the results are counterintuitive) rather than wasting many expensive compute cycles on completing the entire simulation run.

4.5.3 Automatic Error Detection and Recovery

SWAGES comprises several levels of error detection and recovery. At the infrastructure level, ADE provides component supervision, component monitoring, and component restarting in case components fail or crash (for core SWAGES components as well as custom simulation components). Moreover, at the SWAGES level, simulation monitoring, supervision, checkpointing, and restarts guarantee that (eventually) scheduled simulations will finish. This includes mechanisms for SWAGES itself to become "dormant" if no host is available (i.e., the scheduler will save its state to disk and schedule a special shell script for execution that will be able to resume the system). At the simulation client level, error recovery comprises restarting all parallelized clients if any errors prevent a parallelized simulation from finishing. Finally, advanced notification mechanisms are available at each level to inform operators or users about the state of the system at any given time (e.g., through the ADE-GUI or through a Web-based user interface).

4.5.4 SWAGES Compared to Other Frameworks

SWAGES shares several features with other grid middleware systems such as Berkley Open Infrastructure for Network Computing (BOINC), Condor, or QosCosGrid. For example, BOINC (Anderson, 2004) also supports the distribution of parameters and input files from a centralized server to client applications running on remote hosts. Architecturally, there is a similarity between the BOINC manager application and a SWAGES SimClient in that they both broker communication back to the server and execute the client-side simulation. Furthermore, the corresponding BOINC server-side feeder application responsible for providing initialization information to the client simulation can be likened to the role provided by the SWAGES server-side scheduler and *simulation manager*. In SWAGES, the simulation manager fills the additional function of overseeing all communication with its associated SimClient, whereas in BOINC, there are a series of independent services responsible for these interactions (i.e., the respective feeders, validators, and assimilators). Also, similar to other batch submission systems, SWAGES handles distribution of work across a series of available nodes, and like Condor (Bent, 2005) and BOINC, SWAGES is sensitive to the usage availability of the processing host. SWAGES also shares features with QosCosGrid ProActive. Both are Java-based grid middleware systems and support the parallel distribution of simulation (Kravtsov et al., 2008). Additionally, the ability of QosCosGrid to utilize systems crossing security domains (Choppy et al., 2009) is shared by SWAGES. SWAGES can automatically establish a secure shell tunnel for systems behind firewalls and permits the use of either shared or system-unique access credentials when establishing these connections. Furthermore, both systems support an XML markup structure for job submission but differ in their implementation: QosCosGrid's QCG Job Profile supports a broader range of simulation execution types, while SWAGES is a targeted system for large-scale experimentation and therefore has specialized features to support its workflow.

However, unlike most other popular grid systems, SWAGES supports some features not commonly found elsewhere. SWAGES has been tailored to natively support experimentation through integrated parameter sweep mechanisms. This feature, coupled with a Web-based submission capability, makes SWAGES very easy for a modeler to use. Furthermore, SWAGES supports the unique capability of being able to dynamically parallelize a multiagent simulation at the client level across a series of hosts. While some extensions to existing simulations add support for distributed computing (e.g., HLA-RePast) (Minson and Theodoropoulos, 2004), which uses high-level architecture (HLA) to distribute simulations based on the RePast toolkit (Collier, 2001), the distribution is not automatic and does not provide advanced distributed discrete-event scheduling that is found in SWAGES. Furthermore, SWAGES modelers do not have to include any provisions for parallelization in their code. Simply adding the keyword "parallelize" to the experimental setup definition is sufficient for SWAGES to attempt parallelization of simulations whenever possible based on the available computational resources.

SWAGES's fine-grained parallelization and asynchronous scheduling can lead to a much better use of a large array of computational resources when individual simulations are extremely computationally intensive.

4.6 CONCLUSIONS

In this chapter, we provided an overview of the latest version of SWAGES. Many attributes of SWAGES make it an easy system to install and use from a modeler's perspective:

- No preinstalled components are required (only secure shell access).
- No fixed pool of compute nodes is required (SWAGES dynamically adjusts to the pool of available hosts in the grid environment).
- Simulation supervision and recovery mechanisms are included.
- Simulation experiments can be easily defined through a Web-based user interface, which also allows the modeler to look at results and data visualizations as data become available.

And since SWAGES is now implemented in ADE, it provides an open interface that facilitates easy interconnections with other platforms and components, and thus enables easy extensions of SWAGES functionality. In particular, it inherits the flexibility of ADE to use any other component implemented in ADE. For example, existing natural language components (developed in the context of a distributed robotic architecture, DIARC) could be used for user interactions, or action script interpreters for controlling physical agents could be used as scripting engines. Moreover, SWAGES can be easily extended by adding new ADE components based on the standard interface mechanisms provided by ADE (even closed-source commercial off-the-shelf software could be wrapped as ADE components and integrated into the system). Finally, the new distributed architecture allows SWAGES to manage large-scale grid simulations that are not possible for a system with a monolithic single-server configuration.

REFERENCES

D. P. Anderson. BOINC: A system for public-resource computing and storage. In *GRID'04: Proc. of the 5th IEEE/ACM Int'l Workshop on Grid Computing*, pp. 4–10. IEEE Computer Society, 2004.

J. R. Anderson, D. Bothell, M. D. Byrne, et al. An integrated theory of the mind. *Psychological Review*, 111:1036–1060, 2004.

J. Bent. Data-driven batch scheduling. PhD thesis, University of Wisconsin, Madison, 2005.

T. Boyer, M. Scheutz, and B. Bertenthal. Dissociating ideomotor and spatial compatibility. In *Proc. of the 31st Annual Conference of Cognitive Science*, 2009.

M. Charlot, G. de Fabritis, A. L. Lomana, et al. The QosCosGrid project: Quasi-opportunistic supercomputing for complex systems simulations. Description of a general framework from different types of applications. In *Proc. of IBERGRID 2007*, Santiago de Compostela, Spain, 2007.

C. Choppy, O. Bertrand, and P. Carle. Coloured Petri nets for chronicle recognition. In F. Kordon and Y. Kermarrec, editors, *Ada-Europe'09 Proc. of the 14th Ada-Europe Int'l Conference on Reliable Software Technologies*, pp. 266–281, Berlin and Heidelberg: Springer-Verlag, 2009.

N. Collier. RePast: An extensible framework for agent simulation. *Natural Resources and Environmental Issues*, 8:17–21, 2001.

D. Comer. The ubiquitous B-tree. *ACM Computing Surveys*, 11(2):121–137, 1979.

R. T. Fielding and R. N. Taylor. Principled design of modern Web architecture. *ACM Transactions on Internet Technology (TOIT)*, 2(2):115–150, 2002.

K. A. Gluck, M. Scheutz, G. Gunzelmann, et al. Combinatorics meets processing power: Large-scale computational resources for BRIMS. In *Proc. of the 16th Conference on Behavior Representation in Modeling and Simulation*, pp. 73–83, 2007.

V. Kravtsov, A. Schuster, D. Carmeli, et al. Grid-enabling complex system applications with QosCosGrid: An architectural perspective. In *Proc. of The Int'l Conference on Grid Computing and Applications (GCA'08)*, pp. 168–174, 2008.

R. Minson and G. Theodoropoulos. Distributing RePast agent-based simulations with HLA. In *Proc. of the 2004 European Simulation Interoperability Workshop*, Edinburgh, UK, 2004.

M. Peschl and M. Scheutz. Explicating the epistemological role of simulation in the development of theories of cognition. In *Proc. of the 17th Int'l Colloquium on Cognitive Science*, pp. 274–280. Institute for Logic, Cognition, Language and Information, The University of the Basque Country, 2001.

M. Scheutz. The evolution of simple affective states in multi-agent environments. In D. Cañamero, editor, *Proc. of AAAI Fall Symposium*, pp. 123–128, Falmouth, MA: AAAI Press, 2001.

M. Scheutz. ADE—steps towards a distributed development and runtime environment for complex robotic agent architectures. *Applied Artificial Intelligence*, 20(4–5):275–304, 2006.

M. Scheutz. Model-based approaches to learning: Using systems models and simulations to improve understanding and problem solving in complex domains. In P. Blumschein, W. Hung, and D. Jonassen, editors, *Artificial Life Simulations–Discovering Agent-Based Models, volume 4 of Modeling and Simulations for Learning and Instruction, Artificial Life Simulations–Discovering Agent-Based Models*, Rotterdam: Sense Publisher, 2008.

M. Scheutz and B. Gibson. Visual attention and the semantics of space: Evidence for two forms of symbolic control. In *Proc. of the 28th Annual Meeting Cognitive Science Society*, Vancouver, British Columbia, Canada, 2006.

M. Scheutz and J. Harris. Adaptive scheduling algorithms for the dynamic distribution and parallel execution of spatial agent-based models. In F. Fernández de Vega and E. Cantú-Paz, editors, *Parallel and Distributed Computational Intelligence, Volume 269 of Studies in Computational Intelligence*, pp. 207–233, Springer, 2010.

M. Scheutz, J. Harris, and S. Boyd. How to pick the right one: Investigating trade-offs among female mate choice strategies in tree frogs. In *Proc. of the Simulation of Adaptive Behavior 2010*, pp. 618–627, 2010.

M. Scheutz and P. Schermerhorn. The role of signaling action tendencies in conflict resolution. *Journal of Artificial Societies and Social Simulation*, 7(1), 2004.

M. Scheutz and P. Schermerhorn. Predicting population dynamics and evolutionary trajectories based on performance evaluations in alife simulations. In *Proc. of 2005 Genetic and Evolutionary Computation Conference*, pp. 35–42, ACM Press, 2005.

M. Scheutz and P. Schermerhorn. Adaptive algorithms for the dynamic distribution and parallel execution of agent-based models. *Journal of Parallel and Distributed Computing*, 66(8):1037–1051, 2006.

M. Scheutz and P. Schermerhorn. The limited utility of communication in simple organisms. In *Artifical Life XI: Proc. of the 11th Int'l Conference*, pp. 521–528, 2008.

M. Scheutz, P. Schermerhorn, and P. Bauer. The utility of heterogeneous swarms of simple UAVs with limited sensory capacity in detection and tracking tasks. In *Proc. of 2005 IEEE Swarm Intelligence Symposium*, pp. 257–264, Pasadena, CA: IEEE Computer Society Press, 2005.

M. Scheutz, P. Schermerhorn, R. Connaughton, et al. SWAGES—an extendable distributed experimentation system for large-scale agent-based alife simulations. In *Proc. of Artificial Life X*, pp. 412–419, 2006.

M. Scheutz, P. Schermerhorn, J. Kramer, et al. First steps toward natural human-like HRI. *Autonomous Robots*, 22(4):411–423, 2007.

A. Sloman and B. Logan. Building cognitively rich agents using the SIM_Agent toolkit. *Communications of the ACM*, 42(3):71–77, 1999.

J. S. Steinman. Discrete-event simulation and the event horizon. In *PADS'94: Proc. of the 8th Workshop on Parallel and Distributed Simulation*, pp. 39–49, New York: ACM Press, 1994.

Chapter **5**

Repast HPC: A Platform for Large-Scale Agent-Based Modeling

Nicholson Collier and Michael North

Argonne National Laboratory, Argonne, IL, USA

5.1 INTRODUCTION

In the last decade, agent-based modeling and simulation (ABMS) has been successfully applied to a variety of domains, demonstrating the potential of this technique to advance science, engineering, and policy analysis (North and Macal, 2007). However, realizing the full potential of ABMS to find break-through results in research areas such as social science and microbial biodiversity will surely require far greater computing capability than is available through current ABMS tools. The Repast for High Performance Computing (Repast HPC) project hopes to realize this potential by developing a next-generation ABMS system explicitly focusing on large-scale distributed computing platforms.

This chapter's contribution is its detailed presentation of the implementation of Repast HPC as a useful and usable framework. Section 5.2 discusses agent simulation in general, providing a context for the more detailed discussion of Repast HPC that follows. Section 5.3 describes the motivation for our

Large-Scale Computing, First Edition. Edited by Werner Dubitzky, Krzysztof Kurowski, Bernhard Schott.
© 2012 John Wiley & Sons, Inc. Published 2012 by John Wiley & Sons, Inc.

work and the limits of current attempts to increase the computational capacity of ABMS toolkits. Section 5.4 discusses the salient aspects of our existing Repast HPC toolkit that guide the development of Repast HPC. In Section 5.5, we explore the parallelization of an ABMS in a more general way, and Section 5.6 describes the current implementation. Section 5.7 introduces an example Repast HPC social science "rumor" application, while Section 5.8 presents a concluding discussion.

5.2   AGENT SIMULATION

ABMS is a method of computing the potential system-level consequences of the behaviors of sets of individuals (North and Macal, 2007). ABMS allows modelers to specify each agent's individual behavioral rules, to describe the circumstances in which the individuals reside, and then to execute the rules to determine possible system results. Agents themselves are individually identifiable components that usually represent decision makers at some level. Agents are often capable of some level of learning or adaptation ranging from a simple parameter adjustment to the use of neural networks or genetic algorithms. For more information on ABMS, see North and Macal (2007).

5.3   MOTIVATION AND RELATED WORK

The growing roster of successful ABMS applications demonstrates the enormous potential of this technique. However, current implementations of ABMS tools do not currently focus on very rich models or high-performance computing platforms. Many recent efforts to increase the computational capacity of ABMS toolkits have focused on either simple parameter sweeps or the use of standardized but performance-limited technologies. In contrast, Repast HPC's focus is on enabling distributed runs over multiple processes. In doing so, Repast HPC enables (1) massive individual runs, that is, runs containing a number of agents sufficient to overwhelm a single process; and (2) runs containing relatively few complex agents where the computational complexity would overwhelm a single process. ABMS studies typically require the execution of many model runs to account for stochastic variation in model outputs as well as to explore the possible range of outcomes. These ensembles of runs are often organized into "embarrassingly parallel" parameter sweeps due to the way that the independence of the individual model runs allows them to be easily distributed over large numbers of processors on high-performance computers. Quite a few groups have developed tools such as Drone[1] and OMD (Koehler and Tivnan, 2005) for performing such parameter sweeps on high-performance computing systems. Unfortunately, many of the most interesting

---

[1] http://drone.sourceforge.net.

agent-based modeling problems, such as microbial biodiversity, the social-aspects of climate change, and others yet to be developed, require extremely large individual model runs rather than large numbers of smaller model runs. The parameter sweep approach does not offer a solution to this need.

The need for extremely large individual model runs has led various research groups to employ existing standards such as high-level architecture (HLA)[2] and various kinds of Web services (Houari and Far, 2005). These widely used standards can be effective for their purposes, but they are not designed to work efficiently on extremely large computing platforms. This ultimately limits their ability to address the need for extremely large individual model runs. Unfortunately, attempts to introduce new distributed ABMS standards, such as the Foundation for Intelligent Physical Agents (FIPA) architecture specifi-cations[3] and KAoS (Bradshaw, 1996), have succumbed to the same problem. However, despite difficulties in efforts to leverage existing standards as well as the more common focus on large numbers of relatively small model runs, a few groups have developed tools and frameworks for distributing large numbers of agents in a single model across a large number of processors. Deissenberg, van der Hoog, and Dawid's EURACE project (Deissenberg and Hoog, 2009) attempts to construct an agent-based model of the European economy with a very large (up to approximately 10^7) population of agents. Agent-based computational economics (ACE) is a fertile area of research, and while it is important to note that the map (i.e., the model) is not (and need not be) territory, larger models are needed in order to explore the complexity and scope of an actual economy.

EURACE will be implemented using the Flexible Large-Scale Agent Modeling Environment (FLAME). FLAME's origins are in the large-scale simulation of biological cell growth. Each agent is modeled as a "Stream X-Machine" (Laycock, 1993). A Stream X-Machine is a finite state machine consisting of a set of finite states, transition functions, and input and output messages. The state of a Stream X-Machine is determined by its internal memory. For example, the memory may contain variables such as an ID, and x and y positions (Deissenberg and Hoog, 2009). Agents communicate by means of messages, and their internal state is modified over time by transition functions that themselves may incorporate input messages and produce output messages. This representation of agents as Stream X-Machines is similar to Repast's representation of agents as objects, the internal state of which is captured by fields, and the behavior of which is modeled by methods (see next discussion). It is unclear, though, how EURACE/FLAME implements topo-logical (network, spatial, etc.) relations between agents other than via internal memory variables. In contrast, Repast uses more flexible contexts and projec-tions (see next discussion).

---

[2] The SISC's IEEE Standard for Modeling and Simulation (M&S) High Level Architecture (HLA)—Framework and Rules.
[3] http://www.fipa.org.

In common with other large-scale agent simulation platforms, including Repast, EURACE leverages the fact that much agent communication is local. By clustering "neighboring" agents on the same process, costly interprocess communication can be reduced. Here "neighbor" refers to agents within each others' sphere of influence. Thus, an agent may be a network neighbor of an agent whose simulated spatial location is much further away. Of course, there are complications with this approach in that it may not always be possible to arrange agents in a process-compatible way. For example, an agent may need to participate in a network whose other members' simulated spatial location is not architecturally local to that agent. Nevertheless, this principle of distributing agents across processes according to neighborhood or sphere of influence is often efficient and is common to agent simulation platforms.

The SimWorld Agent-Based Grid Experimentation System (SWAGES) (Scheutz et al., 2006) platform provides automatic parallelization via its SimWorld component. The automatic parallelization algorithm uses spatial information from the simulation to calculate agents' spheres of influence and to distribute the agents across processes accordingly. However, such automatic parallelization as described in Scheutz et al. (2006) works only with explicitly spatial simulations rather than with other agent environment topologies.

SWAGES itself is composed of both client and server components and runs on both fixed and dynamic clusters of machines. Most of the server components, running on a single host, manage various aspects of the distributed simulation infrastructure, while the component that manages clients most directly represents the remote simulation and stores the shared simulation state. SWAGES provides an integrated platform for distributed agent-based simulation including data collection and so forth, but unlike Repast HPC, SWAGES is not designed to run on massively parallel machines such as IBM's Blue Gene. Agent implementation must be done in the Poplog programming languages (i.e., Pop11, Prolog, ML, Scheme, C-Lisp) rather than in more common languages such as C++ (used by Repast HPC). Kwok and Chow (2002), although not working in a strictly parallel context, emphasize the importance of proximity and "sphere of influence" with respect to load balancing. Their concern is load balancing for distributed multiagent computing. Agents in this case are more or less autonomous actors performing some task on behalf of an end user. For example, agents may be deployed to locate the lowest price of some commodity in a financial market. In this case, the agents can be thought of as "brokers" and the interaction between them as an exchange of "services." Their Comet load balancing algorithm uses a "credit"-based model for managing the distribution of agents among a cluster of machines. Each agent is dynamically assigned a credit score based on a variety of factors, among which are the computational workload and the proximity of the agents with which the scored agent communicates. The idea is that moving agents off of a process that contains the agents with which it communicates will adversely affect the efficiency of the system. Although the algorithmic work of Kwok and Chow (2002) provides avenues of investigation for Repast

HPC's load balancing, it differs significantly. It is not concerned with very large numbers of agents, and its notion of agents and an agent simulation environment obviously differs.

In a more strictly parallel context, Oguara et al. (2006) also utilize the sphere of influence notion, describing an adaptive load management mechanism for managing shared state, the goal of which is to minimize the cost of accessing the shared state while balancing the cost of managing, migrating, and so on, of the shared state. The cost is minimized by decomposing the shared state according to a sphere of influence and by distributing it so that the shared state is as local as possible to the agents that access it most frequently. The benefits of this minimization are balanced relative to the computational load of actually accomplishing it. Oguara et al. (2006) explore the mechanism using the parallel discrete-event simulation–multiagent system (PDES-MAS) framework. PDES-MAS represents agents as an agent logical process. Agent logical processes have both private internal state and public shared state. The shared state is modeled as a set of variables. Any operation on the shared state (i.e., reading or writing) is a time-stamped event. The sphere of influence of an event is the set of shared state variables accessed by that event. Given the time stamp, the frequency of shared state access can be calculated.

PDES-MAS also contains communication logical processes that manage the shared state. Communication logical processes are organized as a tree whose branches terminate in the agent logical processes. The shared state is distributed and redistributed among the communication logical process tree nodes as the shared state variable migration algorithm is run. When an agent accesses the shared state, that request is routed through the tree and is processed by the communication logical process nodes. The history of this access is recorded and the shared state migration algorithm evaluates the cost of access in order to determine when and how the shared state should be migrated. Additional work is done to avoid bottlenecks (i.e., concentrating too much shared state in too few communication logical processes) as well as to balance the computational cost of migration.

As compelling as the contribution of Oguara et al. (2006) is, it does not provide a full agent simulation package, lacking such things as topology (grid, network, etc.) management, data collection, and so on, that Repast HPC contains. We would in the future like to leverage their work on the efficient management of shared state. The object-oriented nature of Repast HPC's agents makes such externalization of shared state much easier than it might otherwise be. Furthermore, much shared state is the agents' environment as represented by topologies such as grids and networks. Repast HPC externalizes these topologies in separate projection objects, making their management in a distributed context much easier.

Massaioli et al. (2005) take a different approach to the problems of shared state. After experimenting with Message Passing Interface (MPI)-based agent simulations and deciding that the overhead of communicating and synchronizing shared state was too high, Massaioli and his colleagues explore a shared-memory OpenMP-based approach. The majority of their paper (Massaioli et al., 2005) then evaluates the efficiency of various OpenMP idioms in parallelizing typical agent-based simulation structures. In particular, they address optimizing the iteration and evaluation of non-fixed-length structures such as linked lists of pointers. OpenMP is very good at processing the elements of fixed-length arrays in parallel, and Massaioli et al. (2005) investigate how to achieve a similar performance over linked lists of pointers. Although they do highlight issues of sharing and synchronizing states via MPI, we believe a properly architected platform (of which Repast HPC is the first step) can avoid or at least minimize these issues. Indeed, the work of Oguara et al. (2006) illustrates how this might be done. Furthermore, an OpenMP-based approach assumes a shared-memory architecture that may be present within a processor node, but not necessarily across all the nodes. Other works, although not strictly focusing on the large-scale execution of ABMS applications, explore the related area of parallel discrete-event simulation (PDES) (Fujimoto, 1990, 2000). Such work typically focuses on parallel event synchronization where logical processes communicate and synchronize with each other through time-stamped events. Three approaches are commonly used: conservative, optimistic, and mixed mode (Perumulla, 2006, 2007). The conservative approach blocks the execution of the process, ensuring the correct ordering of simulation events (Chandy and Misra, 1979, 1981). The optimistic approach avoids blocking but must compensate when events have been processed in the incorrect order. The Time Warp operating system is the preeminent example of this approach (Jefferson, 1985; Jefferson et al., 1987). Mixed mode combines the previous two (see, e.g., Perumulla [2005] and Jha and Bagrodia [1994]). The trio of Sharks World papers (Nicol and Riffe, 1990; Presley et al., 1990) explores the application of these approaches to the Sharks World simulation (sharks eating and fish swimming in a toroidal world). The scalability of these approaches to more recent high-performance computing platforms such as IBM's Blue Gene is discussed by Perumulla (2007), where he demonstrates scalability to over 10,000 CPUs.

Perumalla's microkernel for parallel and distributed simulation, known as "μsik" (Perumulla, 2005, 2007) provides efficient PDES time scheduling. The correctness of simulation results requires that events are executed such that "global time-stamp order" execution is preserved. In essence, no process should execute events with time stamps less than its current time stamp, at least without compensation as in the optimistic approach described earlier. Doing this efficiently requires that global parallel synchronization must be fast.

The μsik microkernel (Perumulla, 2005, 2007) optimizes execution efficiency by minimizing the costs of rollback support, using a fast scalable algorithm for global virtual time (GVT) computation, and by minimizing the cost of fossil (i.e., unnecessary state and event history) collection. By memory reduction techniques such as employing reverse computation for rollback rather than state saving, significant improvements were achieved. Perumulla (2007) also illustrates how key algorithms, such as the GVT computation, scale effectively to 10^4 processors without significant inefficiencies. Memory reduction is also helped by μsik's fast logarithmic fossil collection. With respect to event scheduling, Repast HPC uses a distributed discrete-event scheduler that is, at least in the current implementation, tightly synchronized across processes. In this, it has more in common with conservative PDES synchronization algorithms as discussed earlier (Chandy and Misra, 1979, 1981). Peramulla (2005, 2007) illustrates well the complexities and concomitant performance payoffs associated with more sophisticated scheduling. The initial implementation of the Repast HPC scheduler avoids such complexities but leaves room for implementing more advanced parallel scheduling algorithms in the future. Lastly, although μsik does provide sophisticated PDES support, unlike Repast HPC, it does not provide the additional services needed for agent modeling such as topology management, data logging, and so forth.

Like Perumalla (2005, 2007), Bauer et al. (2009) implement and test efficient discrete time scheduling. In particular, they demonstrate scalable performance of a Time Warp-based simulator using reverse computation for rollback. Their results show effective scaling up to 32,768 processors on an IBM Blue Gene/L and up to 65,536 processors on a Blue Gene/P. They attribute the ability to scale to their message/network management algorithms that are tuned to leverage Blue Gene's high-performance asynchronous message capabilities. Their implementation is based on ROSS: Rensselear's Optimistic Simulation System (Carothers et al., 2002).

As in the case of Perumalla, Bauer et al. (2009) focus on PDES optimization and experimentation and unlike our work, they do not support the additional services needed for agent modeling such as topology management, data logging, and so forth. In addition, while they leverage Blue Gene's message capabilities to achieve their efficiencies, Repast HPC aims at more general applicability. Notwithstanding these considerations, we hope to benefit from their work in the future in implementing more sophisticated parallel scheduling algorithms.

Karimabadi et al. (2006) study PDES for selected classes of grid models, in particular, hybrid models of electromagnetism. Their basic strategy is to use an approach similar to lazy program evaluation (Hudak, 1989). In other words, they set up their PDES to only execute an event when the consequences of that event are needed by some other event. Implementing their algorithm using Perumalla's μsik (Perumulla, 2005) engine, they report a greater than 100-fold performance improvement over traditional PDES. Unfortunately, their algorithm is limited in the kinds of events it can support. Furthermore, like Perumalla's μsik (Perumulla, 2005), they do not provide the additional services needed for agent modeling.

Some researchers such as D'Souza and colleagues (D'Souza et al., 2007; Lysenko and D'Souza, 2008) have studied the use of graphical processing units (GPUs) either individually or in large groups. In general, substantial speedups are found, but at the expense of a substantial reduction in modeling flexibility. GPUs are designed to rapidly and repeatedly execute small sets of instructions using limited instruction sets. This allows the underlying GPU hardware to be highly optimized for both speed and production cost. Efficient use of GPUs implies that the executing agents will tend to run the fastest when they have simple behaviors drawn from a short shared list of options.

This chapter's contribution is its detailed presentation of the implementation of Repast HPC, the first complete ABMS platform developed explicitly for large-scale distributed computing systems. Our work on Repast HPC provides an integrated ABMS toolkit targeted at massive individual runs on high-performance computing platforms. In this, it differs from tools that enable distributed parameter sweeps over many individual runs. We leverage the related work done in distributed ABMS and PDES and hope to take more advantage of it in the future, but our focus is on a working integrated toolkit that provides the typical ABMS components specialized for a massively parallel environment. In addition, using standards such as MPI and C++, the code itself is compatible across multiple computing platforms.

Related work and how it compares to this chapter and Repast HPC is summarized next.

Belding (Drone)

- **Citation:** Koehler and Tivnan (2005)
- **Summary:** distributed parameter sweeps of relatively small runs
- **How Repast HPC Differs:** support for very large individual model runs

IEEE, FIPA

- **Citation:** Houari and Far (2005) and Bradshaw (1996)
- **Summary:** standards for model interoperability and distribution
- **How Repast HPC Differs:** actual working system for very large-sized model runs

EURACE

- **Citation:** Deissenberg and Hoog (2009)
- **Summary:** extremely large-scale distributed model of European economy
- **How Repast HPC Differs:** flexible and generic components based on contexts and projections

FLAME

- **Citation:** http://www.flame.ac.uk
- **Summary:** flexible large-scale agent modeling architecture based on Stream X-Machines
- **How Repast HPC Differs:** flexible and generic components based on contexts and projections using standard OO-techniques

SWAGES

- **Citation:** Scheutz et al. (2006)
- **Summary:** distributed simulation systems for large-scale ABMS
- **How Repast HPC Differs:** supports parallelization of more than just spatial simulations

Load Balancing

- **Citation:** Kwok and Chow (2002)
- **Summary:** load balancing for distributed multiagent computing
- **How Repast HPC Differs:** supports very large-sized distribution and employs a different although related notion of an agent

PDES-MAS

- **Citation:** Oguara et al. (2006)
- **Summary:** explores adaptive load management for managing shared state
- **How Repast HPC Differs:** less sophisticated shared state management but working support for all the typical agent simulation components

OpenMP for Agent-Based Simulation

- **Citation:** Massaioli et al. (2005)
- **Summary:** explores and evaluates the use of OpenMP in parallelizing agent-based simulation structures
- **How Repast HPC Differs:** uses MPI and provides a full set of working ABMS components.

PDES Scheduling in Sharks World

- **Citation:** Bagrodia and Liao (1990), Presley et al. (1990), and Nicol and Riffe (1990)
- **Summary:** explore PDES scheduling (conservative, optimistic, and mixed mode) in the context of the Sharks World simulation
- **How Repast HPC Differs:** simpler scheduling but *working* support for all the typical agent simulation components

μsik PDES Kernel

- **Citation:** Perumulla (2005)
- **Summary:** describes how the μsik PDES kernel provides efficient event scheduling

- **How Repast HPC Differs:** simpler scheduling but working support for all the typical agent simulation components

μsik Scalability

- **Citation:** Perumulla (2007)
- **Summary:** describes how μsik scales to very large sizes
- **How Repast HPC Differs:** simpler scheduling but working support for all the typical agent simulation components on high-performance computing machines.

Lazy PDES Event Execution

- **Citation:** Karimabadi et al. (2006)
- **Summary:** explores the use of lazy PDES event execution using the μsik kernel
- **How Repast HPC Differs:** simpler scheduling but working support for all the typical agent simulation components

GPUs for ABMS

- **Citation:** D'Souza et al. (2007) and Lysenko and D'Souza (2008)
- **Summary:** explores the use of GPU and GPU-based programming as a platform for ABMS
- **How Repast HPC Differs:** much more flexible programming model

5.4 FROM REPAST S TO REPAST HPC

Repast is a widely used ABMS toolkit.[4] Multiple versions of the toolkit have been created, the latest of which is Repast Simphony for Java (Repast SJ). Repast SJ is a pure Java ABMS platform whose architectural design is based on central principles important to agent modeling. These principles combine findings from many years of ABMS toolkit development and from experience applying the ABMS toolkits to specific applications. More details on Repast Simphony are reported by North et al. (2007).

The parallel C++ implementation of Repast (Repast HPC) attempts to preserve the salient features of Repast SJ while adapting them for parallel computation in C++. More specifically,

- agents are naturally represented as objects (in the object-oriented programming sense);

---

[4] http://repast.sourceforge.net.

- scheduling is flexible and dynamic; and
- implementation should be driven by modeling rather than framework requirements; model components (e.g., agents) should be "plain old objects" as far as possible.

These requirements are considered in detail in the remainder of this section.

5.4.1 Agents as Objects

An agent's internal state (e.g., its age and wealth) is easily represented in an object's fields while the agent's behavior (e.g., eating, aging, and acquiring and spending wealth) is modeled using an object's methods. Implementing agents as objects is most easily done using an object-oriented language. Consequently, Repast HPC is implemented in C++. Using C++ also allows us to take advantage of the Standard Template Library (STL) as well as sophisticated third-party libraries such as Boost.[5]

5.4.2 Scheduling

Repast Simphony simulations are driven by a discrete-event scheduler. Events themselves are scheduled to occur at a particular tick. Ticks do not necessarily represent clock time but rather the priority of its associated event. Ticks determine the order in which events occur with respect to each other. For example, if event *A* is scheduled at *tick* 3 and event *B* at *tick* 6, event *A* will occur before event *B*. Assuming nothing is scheduled at the intervening ticks, *A* will be immediately followed by *B*. There is no inherent notion of *B* occurring after a duration of three ticks. Of course, ticks can and are often given some temporal significance through the model implementation. A traffic simulation, for example, may move the traffic forward 30 seconds for each tick. A floating point tick, together with the ability to order the priority of events scheduled for the same tick, provides for flexible scheduling. In addition, events can be scheduled dynamically on the fly such that the execution of an event may schedule further events at that same tick, or some future tick.

5.4.3 Modeling

Repast Simphony attempts to separate framework concerns from modeling concerns by eliminating implementation requirements such as interfaces or abstract base classes on the model side. Agents need not implement any interfaces or extend any parent classes in order to participate in the framework. Historically, agent-based models have maintained a tight coupling between individuals, their behaviors, and the space in which they interact. As a result,

---

[5] http://www.boost.org.

many models have been designed in a way that limits their ability to express behaviors and interactions. For example, an agent may have to implement a network node interface in order to participate as a node in a network, or a grid element interface in order to participate in a grid, or both to participate in both a network and a grid. Such coupling of the agent to the space in which it participates reduces flexibility and reuse, adds an undue implementation burden, and diverts focus from the actual model.

Repast Simphony's approach avoids these problems and encourages flexibility and reusability of components and models through the use of contexts and projections. The context is a simple container based on set semantics. Any type of object can be put into a context, with the simple caveat that only one instance of any given object can be contained by the context. From a modeling perspective, the context represents a population of agents. The agents in a context are the population of a model. However, the context does not inherently provide any mechanism for interaction between the agents. Projections take the population as defined in a context and impose a new structure on it. The structure defines and imposes relationships on the population using the semantics defined in the projection. Projections have a many-to-one relationship with contexts. Each context can have an arbitrary number of projections associated with it. Projections are designed to be agnostic with respect to the agents to which projections provide structure. Actual projections are such things as a network that allows agents to form network-type relations with each other or a grid-type space where each agent is located in a matrix-type space (Howe et al., 2006).

Repast HPC attempts to preserve these features and principles as far as possible. Repast HPC has additional goals, driven by ease of use, namely,

- Users of the framework should be able to work with projections and agents as such, rather than more typical parallel friendly types: arrays, primitives, and so forth.
- Similarly, the details of parallelization (e.g., MPI-related calls) should be hidden from the client programmer (i.e., the user) as much as possible.

5.5   PARALLELISM

In parallelizing Repast for large-scale distributed platforms, our focus is on parallelizing the simulation as a whole rather than enabling the parallelization of individual agent behaviors across processors.[6] The expectation, then, is that the typical simulation will contain either relatively few heavy, computationally complex agents per process or a multitude of lighter-weight agents per process. Processes communicate, share agents, and so forth using the MPI.

---

[6] The latter could be done, of course, either through a model-specific scheme or by conceptualizing the agent behavior as kind of pseudosimulation and by using Repast HPC's functionality.

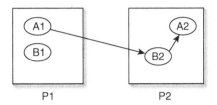

Figure 5.1 *Agent A1 with a network link across processes to agent B2.*

Each individual process is responsible for some number of agents local to that process. These agents reside in the memory associated with that process and the process executes the code that represents these local agents' behavior. In terms of the Repast HPC components, each process has a scheduler, a context and the projections associated with that context. At each tick, a process scheduler will execute the events scheduled for that current tick, typically some behavior of the agents in the context that makes use of a projection. For example, each agent in a context might examine its network neighbors and might take some action based on those links, or each might examine its grid surroundings and move somewhere on the grid based on what it finds.

Processes do not exist in isolation. The schedule is synchronized across processes and data production and collection can be aggregated across all processes. More importantly, copies of nonlocal agents (i.e., agents whose own behavior is initiated by another process schedule) may reside on a process. These nonlocal agents may be used by the local agents as part of their behavior but are not under the control of the executing process. For example, agents can participate in a shared network that spans processes such that a link may exist between local and nonlocal processes.

Figure 5.1 shows a link between agent A1 and agent B2. A1 resides on process P1 and B2 on process P2. However, this cross-process link does not actually exist as such. A copy of B2 exists on P1 and A1 links with that. A1 can then query its network neighbors when its behavior is executed.

When copies of agents can be shared across processes, it is then necessary to synchronize the copies with the original agents. This applies both to an agent's state (its properties and so forth) but also to its status (whether it is "alive," "dead," or has migrated among processes). If, for example, in the course of its behavior, an agent "dies," then it should be removed from the simulation as a whole. Such synchronization requires a way to uniquely identify an agent without which there is no way to map copies to their originals. The `AgentId` serves this function.

```
class AgentId {
    . . .
    int id_, startProc_, agentType_, currentProc_;
```

```
std::size_t hash;
  . . .
};
```

An `AgentId` is immutable with respect to its `id_`, `startProc_` as well as `agentType_` fields. The `agentType_` field should identify the type of agent. For example, in a predator prey model, a constant value of 0 might identify the predator agents, and a value of 1 might identify the prey agents. The field `startProc_` identifies the process rank on which the agent with this `AgentId` was created. Lastly, `id_` is used to distinguish between agents of the same type created on the same process. Prey 2 created on process 3 is distinguished from prey 3 created on process 3, for example. Taken together, these three values should uniquely identify an agent across the entire simulation. The hash field is also immutable and is calculated based on the `id_`, `startProc_` and `agentType_` fields and is used as a key value when using agent objects in hash-based maps. The last field, `currentProc_`, is the process where the agent currently resides. Initially, this is the same as the agent's starting process (`startProc_`) but may change if the agent moves between processes.

Every agent in the simulation must implement a `getAgentId` method that returns the `AgentId` for that agent. Using the `AgentId`, processes can request specific agents from the process where the agent resides and can then update their local copy with the canonical version. Similarly, if an agent dies and is removed from the simulation, the `AgentId` of the dead agent is sent to the relevant processes, allowing them to remove the agent from their contexts, projections, and so on. The actual process of requesting and synchronizing agents is handled by the `RepastProcess` component that is further discussed in greater detail.

5.6 IMPLEMENTATION

To implement its core parallelism, Repast HPC uses the Boost.MPI library[7] and the MPI C++ API where necessary. Boost.MPI applies a more modern C++ interface to the MPI C and C++ API. Boost.MPI provides better support for modern C++ development styles and includes complete support for user-defined data types and C++ Standard Library types. The use of Boost.MPI is also an advantage for the client programmer, especially in the creation of MPI data types for user-defined data types such as agents. This helps to keep the focus on modeling concerns rather than on framework requirements.

The remainder of this section discusses the core components of Repast HPC and their implementation in more detail.

---

[7] http://www.boost.org/docs/libs/1_39/doc/html/mpi.html.

5.6.1 Context

The context component is a simple container based on set semantics that encapsulates an agent population. The equivalence of elements is determined by the `AgentId`. Each process has at least one context to which local and nonlocal agents are added and removed. The context component provides iterator implementations that can be used to iterate over all the agents in the context or only those that are local to the current process. Projections (`Grids` and `Networks`) are also contained by the context such that any agent added to a context becomes a member of the associated projections (e.g., occupying a grid cell or becoming a node in a network).

5.6.2 RepastProcess

The `RepastProcess` component encapsulates a process and manages interprocess communication and simulation state synchronization. Implemented as a singleton together with related template functions, a single `RepastProcess` exists per process. The `RepastProcess` exposes such data as a process rank and the total number of processes. Using the `RepastProcess` component, a client programmer can request copies of agents from other processes and can synchronize the state and status (i.e., alive, dead, or migrated) of requested agents.

```
AgentRequest request(myRank);
// get 10 random agents from other process
for (int i = 0; i < 10; i++) {
  AgentId id = createRandomId();
  request.addRequest(id);
}
repast::requestAgents<AgentContent>(provider, receiver);
```

This piece of code illustrates a request for agents. Requests are made using the `AgentRequest` object to which the IDs of the requested agents are added. This request is then passed to a template function, `requestAgents`. The function is specialized to provide and receive the type `AgentContent`. The provider and receiver, which are themselves type parameters, are responsible for providing the content to the requesting process and for receiving the requested content. The user is responsible for coding the provider and the receiver, but this requires only the implementation of a single method each. The provider typically retrieves the content from agents in the context using the IDs of the requested agents, while the receiver typically creates the agents from the requested content and then adds them to the context.[8]

---

[8] Future work should capture this typical case such that the client programmer would only have to provide the provider and receiver if he or she needs more than simple add and *retrieve* from the context.

The relevant dynamic content or state of an agent can typically be described by a `struct` or class of primitive types. The client programmer specializes the `requestAgents` method for this `struct` or class (e.g., the `AgentContent` type in the above-mentioned example). More importantly, Boost.MPI transparently constructs the necessary MPI data types for such a `struct` or `class` and can efficiently pass (`send`, `recv`, `broadcast`, etc.) them among processes. The amount of work that the client programmer needs to do is fairly trivial; most of the work is done by the `RepastProcess` component and Boost.MPI.

Implementation-wise, the sending and receiving of agent content using the `RepastProcess` component works as follows: Nonroot processes send their requests to the root process (the process with rank 0). The root process sorts these requests as well as any of its own by the containing process, that is, by the process on which the requested agent resides. For example, if P1 requests Agent A from P2, P0 requests B from P2, and P3 requests C from P1, then P0 (the root process) sorts these three requests into two lists: the agents requested from P2 (A and B) and the agents requested from P1 (C). The root process then sends these request lists to the process that can fulfill them. These processes then fulfill the requests by extracting the content of the requested agents and then sending that content to those processes that initiated the request. Throughout this mechanism, `RepastProcess` tracks the agent content that it is importing from other processes and the agent content it is exporting to other processes. Synchronization of state and status is then easily done by sending and receiving the canonical state and status changes between the associated importing and exporting process pairs.

5.6.3 Scheduler

Repast HPC uses a distributed discrete-event scheduler that is, at least in the current implementation, tightly synchronized across processes. In this, it has more in common with conservative PDES synchronization algorithms as discussed earlier (Chandy and Misra, 1979, 1981). Earlier work on parallel simulation demonstrates the effectiveness as well as the complexities of optimistic and mixed-mode scheduling (Jefferson et al., 1987; Jha and Bagrodia, 1994; Perumulla, 2005; Perumulla, 2007). The initial implementation of the Repast HPC scheduler avoids these complexities and leaves room for implementing more advanced parallel scheduling algorithms in the future.

```
MethodFunctor<NetworkSim>*
    mf   =   new   MethodFunctor<NetworkSim>   (this,
&NetworkSim::step); scheduler.scheduleEvent(1, 1, mf);
. . .
scheduler.scheduleStop(100);
```

The scheduler schedules `MethodFunctors`, that is, method calls on objects, using the tick-based scheme described in Section 5.4. `MethodFunctors` can be scheduled to execute once at a particular tick or by starting at a particular tick and then repeating at a regular interval. The scheduler also has additional methods for scheduling simulation termination as well as methods for scheduling events that should occur post-termination. These last events are particularly useful for finishing data collection, cleaning up open resources, and so forth. The code mentioned earlier illustrates a typical piece of scheduling code. The step method on a `NetworkSim` object is scheduled to execute at tick 1 and then every tick thereafter, and the simulation itself is scheduled to terminate at tick 100.1.

The scheduler runs in a loop until the stop event is executed. At each iteration of the loop, the scheduler determines the next tick that should be executed, pops the events for that tick-off of a priority queue, and executes any `MethodFunctors` associated with those events. The scheduler synchronizes across processes when it determines the next tick to execute. Each individual scheduler on each process passes the next tick it will execute to an "all_reduce" call and receives the global minimum next tick as a result. If this global minimum is equal to the local next tick, then the local scheduler executes. Synchronization is thus twofold: (1) The "all_reduce" call is a barrier preventing individual processes from getting too far ahead or too far behind; (2) by only executing the global minimum tick, we insure that all processes are executing the same tick. This type of synchronization is, of course, restrictive and implies that each process is under a similar load and that the coherency of the simulation as a whole benefits from such synchronization. Future work will examine relaxing the synchronization through checkpointing, rollbacks, and so forth. This type of tight synchronization, however, does fit reasonably well with many types of currently implemented agent simulations.

5.6.4 Distributed Network

The distributed network component implements a network projection that is distributed over the processes in the simulation. Each process is responsible for the part of the network in which its local agents participate as nodes. However, these local agents may have network links to nonlocal agents (or rather copies of them as described earlier) and through them to those parts of the network on other processes. The distributed network provides typical ABMS network functionality: retrieving network node successors and predecessors, creating and removing edges, and so forth.

The design of the distributed network is guided by the notion that each process should only see as much of the network as necessary for the behavior of the local agents. For example, if an agent's behavior incorporates only its immediate network neighbors, then only these neighbors need be present on the local process. There is no need for a larger view of the network incorporating neighbors of neighbors, and other features. The extent of the larger network that is visible to a process is thus configurable by the user. In particular, the distributed network includes methods for creating complementary edges across processes as well as for mirroring the relevant linked parts of the larger network into an individual process.

The creation of complementary edges works much like the sharing of agents previously described. For example, Process 1 (P1) creates an edge between A1 and B2 where B2 is a copy of an agent from Process 2 (P2). Initially, P2 has no knowledge of this edge until P1 informs P2, and P2 creates the complementary edge (A1 B2) on its copy of the network. As part of edge creation, the distributed network tracks those edges that require a complement to be created and the process on which they should be created. A list of these processes is sent to the root process. The root process sorts these lists such that a list of processes from which to expect a complementary edge can be sent to each process. Each process then asynchronously sends and receives the complementary edges. These edges are then incorporated into the local copy of the network. Copies of nonlocal agents may be received together with the edge (e.g., P2 receives a copy of A1, mentioned earlier), and these copies are then added to the local context as well as to the RepastProcess' relevant import and export lists.

Acquiring additional parts of the network works in much the same way. Edges are sent and received, although the initial request is, of course, not for complementary edges but for additional nodes in the network.

5.6.5 Distributed Grid

The distributed grid projection implements a discrete grid where agents are located in a matrix-type space. It implements typical ABMS grid functionality: moving agents around the grid, retrieving grid occupants at particular locations, getting neighbors, and so forth. The grid is distributed over all processes, and each process is responsible for some particular subsection of the grid (see Fig. 5.2). The agents in that particular subsection are local to that process. Grids are typically used to define an interaction topology between agents such that agents in nearby cells interact with each other. In order to accommodate this usage, a buffer can be specified during grid creation. The buffer will contain nonlocal neighbors, that is, agents in neighboring grid subsections. Grid creation leverages MPI_Cart_create so that processes managing neighboring grid subsections can communicate efficiently.

The distributed grid also allows agents to move out of one grid subsection and into another. This entails migrating the agent from one process to another in order to minimize the nonlocal knowledge (and thus the synchronization overhead) that each process requires. The distributed grid works in conjunction with a RepastProcess to achieve this interprocess agent movement. Note that this can become quite involved if the agent or copies of the agent are referenced in other processes.

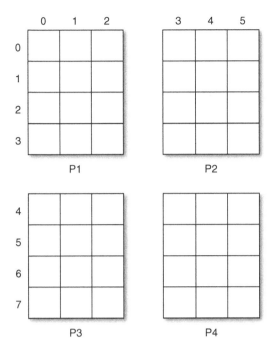

Figure 5.2 *Distributed grid. P1 manages grid cells (0, 0)–(2, 3), P2 manages (3, 0)–(5, 3), and so on.*

5.6.6 Data Collection and Logging

Repast HPC implements two kinds of data collection: aggregate and nonaggregate. Aggregate data collection uses MPI's reduce functionality to aggregate data across processes and writes the result into a single file using Network Common Data Form (NetCDF[9]). The details of MPI and NetCDF are invisible to the user, however. The user specifies numeric data sources and the reduction operation to apply to the data produced from those sources. These data sources are then added to a `DataSet` object, associating the data sources with identifying names. The resulting data are two dimensional: Each column corresponds to a data source and each row to the time (tick) at which the data were recorded. The data itself are retrieved from implementations of a templated interface with a single `getData` method. The type (int or double) of the data returned is determined by the template specialization. The actual recording and writing of data is done through scheduled events. Boost.MPI is particularly helpful here in that it allows the use of the STL C++ algorithms as reduction operations and makes it relatively easy to create custom operations.

Nonaggregate data collection works in much same the way, except that no reduce operation is specified. The resulting data are three dimensional with the process rank acting as the third dimension.

---

[9] http://unidata.ucar.edu/software/netcdf.

Repast HPC also implements a more generic logging framework along the lines of the popular Java Log4J library.[10] The user defines appenders and loggers. Appenders determine where logged data can be written: `stderr`, `stdout`, or to a file(s). A logger combines one or more appenders and a logging level. Any data written to a logger with a level greater than or equal to the logger's log level will then be time-stamped and written to that logger's appenders. Repast HPC creates a default logger for each process that can be used to systematically log debug info, errors, and so forth.

5.6.7 Random Number Generation and Properties

Repast HPC provides centralized random number generation via a random singleton class. Random leverages Boost's random number library, using its implementation of the Mersenne Twister (Matsumoto and Nishimura, 1998) for its pseudorandom number generator. Uniform, triangle, Cauchy, exponential, normal, and log normal distributions are available. Random is not distributed across processes, but rather, individual instances execute on individual processes. Our experience with Repast and agent simulation in general has illustrated the usefulness of a singleton-based random number implementation, especially with respect to replicating results. Repast HPC thus follows this model.

The random seed can be set either in the code itself or by using a simple properties file format. The canonical Repast HPC executable expects a `-config` option that names a properties file. Specific random number properties in that file are then used in the initialization of the random class. A properties file has a simple *key = value* format. An example of random number-related properties is

```
random.seed = 1
distribution.uni_0_4 = double_uniform, 0, 4
distribution.tri = triangle, 2, 5, 10
```

The `random.seed` property specifies the seed for the generator. If no seed is specified, then the current calendar time is used (i.e., the result of `std::time(0)`). Distributions can be created and labeled by specifying a distribution property. The keys for such properties begin with `distribution` followed by a "." and then a name used to identify the distribution for later use. The value of the property describes the type of the distribution and any parameters it might take. For example, the piece of code starting with `random.seed = 1` creates a distribution named "tri." It is a triangle distribution with a lower bound of 2, a most likely value of 5, and an upper bound of 10. This distribution can be used in the model by retrieving it by name ("tri") from the `Random` class.

---

[10] http://logging.apache.org.

In addition to providing easy random number configuration, properties and properties files also provide support for arbitrary key/value pairs. These can be parsed from the properties file itself or added at run time to a `Properties` object. The `Properties` class provides this functionality. The canonical `main` function used by a Repast HPC model will pass the properties file to the `Properties` class constructor and will then make this `Properties` object available to the model itself. Model parameters will typically be set in properties files, making it trivial to change a model's parameters without recompilation.

5.7 EXAMPLE APPLICATION: RUMOR SPREADING

As a benchmark application and equally importantly to determine, if only informally, Repast HPC's ease of use and how well modeling concerns are separated from framework concerns, we have written a simple rumor-spreading application. The application models the spread of a rumor through a networked population. The simulation proceeds as follows. As part of initialization, some numbers of nodes are marked as rumor spreaders. At each iteration of the simulation, a random draw is made to determine if the neighbors of any rumor-spreading nodes have received the rumor. This draw is performed once for each neighbor. After all of the neighbors that can receive the rumor have been processed, the total number of nodes that have received the rumor is recorded. We hope that this initially simple model can become the basis for more thorough investigations of network topologies in related models, such as the spread of trends through a population, the role of "influencers" or trendsetters in such, and the spread of epidemics in general. The potential value of using such virtual "worlds" to generate data for testing analysis techniques is documented by Zurell et al. (2010). In particular, they describe the role of virtual data in evaluating methods for data sampling, analysis, and modeling methods with respect to ecology and ecological models.

The network itself is created using the KE model for growing networks. Klemm and Eguíluz (2002) demonstrated a model for creating networks that display the typical features of real-world networks: power law distribution of degree and linear preferential attachment of new links. More importantly, the KE model creates networks with degree distribution similar to the preferential attachment model of Barabási and Albert (1999) but also includes other network topology features more like those found in real computer networks.

The KE model generates the network as follows. As an initial condition, there is a fully connected network of m active nodes. A new node, i, is added to the network. An outgoing link is created between i and each of the active nodes. Each node j of the m active nodes thus receives exactly one incoming link. The new node i is added to the set of active nodes. One of the currently active nodes is then removed from the set of active nodes. The probability that a node is made inactive is inversely proportional to its current degree, k_i:

$$P(k_i) = \frac{1}{k_i \sum_j \frac{1}{k_j}}. \tag{5.1}$$

This is then repeated for each new node, i, that is to be added to the network (Briesemeister et al., 1998; Klemm and Eguíluz, 2002).

We have adapted this to the distributed context by sharing n number of nodes between processes. Prior to network creation, each process requests n nodes at random from 4 "adjacent" processes. Adjacency here is determined by a process rank, such that a process with rank n is adjacent to ranks $n - 1$, $n - 2, n + 1$, and $n + 2$. When adjacent ranks are less than 0 or greater than the number of processes—1, the adjacent ranks are determined by wrapping either to the end or to the beginning. For example, given a process total of 128, process 0 is adjacent to 126, 127, 1, and 2. After importing the additional nodes, each process then builds a KE-model network using these shared nodes, together with nodes local to that process. Once the network has been built on each process, complementary edges are generated such that the edges created between shared and local nodes are then themselves shared across the relevant processes. For example, an edge is created on process one (P1) creates an edge between two nodes where the first node resides on process two (P2). Complementary edge creation will then create the corresponding edges between the corresponding nodes on P2. The resulting network is intended to roughly model more densely connected clusters of friends integrated into larger clusters of clusters.

As the simulation runs, each process maintains a list of the nodes that have currently received the rumor (so-called rumored nodes). This list includes both local and nonlocal shared nodes. At each iteration of the simulation, we iterate through this list and attempt to pass on the rumor, via random draws, to the neighbors of these rumored nodes. These neighbors, however, must be local to the executing process in order for the rumor attempt to be made. This avoids any race conditions with respect to whether or not a node has received the rumor and the necessity for any potentially time-consuming resolution of conflicting state. The state of shared nodes is updated after processing the neighbors. Any change in the rumor state of nonlocal shared nodes is thus communicated among processes and the list of rumored nodes is updated to include any newly "rumored" nonlocal shared nodes. In this way, the integrity of the network with respect to spreading the rumor via cross-process links is preserved.

The simulation has the following parameters:

1. The name of the distribution to use when determining if a node receives the rumor, via a random draw
2. The probability of a node receiving the rumor
3. The initial node count

4. The value of m to use in creating the KE-model network

5. The value of n, specifying how many shared nodes should be used in the initial network, as described earlier

6. The initial number of rumored nodes—this value can be a constant or the name of a distribution from which the value is drawn

These parameters are contained in a Repast HPC properties file together with named distributions and an initial random seed. The seed is shared among all processes, allowing for repeatable results.

The implementation of the model, of course, uses the various components described previously, the `RepastProcess`, the distributed network, and so forth. Consequently, the nodes are modeled agents and the edges themselves are instances of the `RepastEdge` class. No doubt this thoroughly object-oriented implementation imposes some overhead, but it does have significant advantages with respect to expanding and improving the model. For example, if the spread of a rumor between two nodes becomes dependent on the strength of the tie between them or on some property of the nodes themselves, then the changes to the code are easy to implement.

The code itself was developed on Mac OSX using Open MPI. Benchmark runs and additional testing were performed on an IBM Blue Gene/P 1028 processor system hosted by Argonne National Laboratory's Leadership Computing Facility. Blue Gene's native compiler, bgxlc++, was used to compile the relevant parts of Boost and the Repast HPC code.

5.7.1 Performance Results

The model was run with the following parameters. The probability of a node receiving a rumor was set to 0.01, and the draw was made from a uniform distribution (0, 1). Ten nodes were marked as initial rumored nodes. The number of nodes n shared between adjacent processes was set to 40, and the KE model m parameter was set to 10. Each run iterated for 200 ticks (iterations). Run times were measured for both initialization (agent creation, network setup, initial sharing of node, and edges) as well as the time to complete 200 iterations.

Figure 5.3 illustrates the total run time (initialization plus the time to complete 200 iterations) for varying network node counts grouped by process count. We can see a trade-off between the benefit of running smaller numbers of agents (network nodes) on each process and the cost of interprocess communication. The total run time for a 100-k node network on 256 processes is less than all the other process counts for the same sized network. As the network node count increases, the advantages of running smaller numbers of nodes on each process increases and runs on a larger number of processes become progressively faster. The lack of perfect weak scaling is obvious.

This trade-off is obviously model dependent. Our sample model is not computationally intensive, and thus the advantages of a larger number of

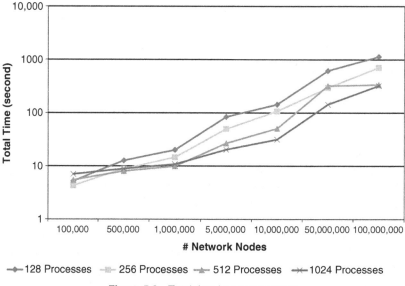

Figure 5.3   *Total time by process count.*

processors do not appear at lower numbers of agents. Models with more computationally "thick" agents would presumably see a benefit with lower numbers of agents and equivalent numbers of processes. Similarly, models that require less interprocess communication would obviously be faster as well.

Figures 5.4 and 5.5 and Table 5.1 illustrate that the cause of the trade-off lies in the initialization of the model and less in its actual execution, that is, in the iteration and execution of the scheduled events. The initialization time for 1024 process runs is nearly 4.50 times slower when running 100-k agents than the 128 process run, even though each individual process is initializing 8× less agents. As mentioned earlier, initialization consists in part of sharing agents between neighboring processes. In terms of the Repast HPC components, this sharing is initiated using `AgentRequests` as described in Section 5.6.2. `AgentRequests` are coordinated by a root process (process rank 0). This root process sequentially receives all the requests sent by all the requesting nonroot processes. The time spent in this send and receive increases as the process count increases: The loop is longer; there is ultimately more interprocess communication and more requests for the root process to coordinate. Future work will investigate potential optimizations to the `AgentRequests` mechanism.

Absent the initialization overhead, the model and the Repast HPC components scale reasonably well. In all but one case, a greater number of processes lead to faster run times.[11] On average, twice as many processors for runs with

---

[11] The exception is the 100-k node run where 256 processes run 0.10 seconds faster than 512.

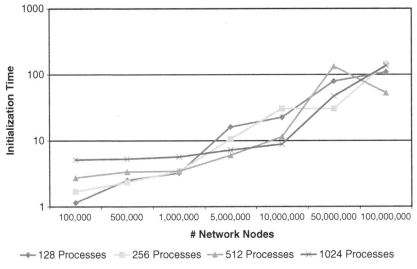

Figure 5.4 *Initialization time by process count.*

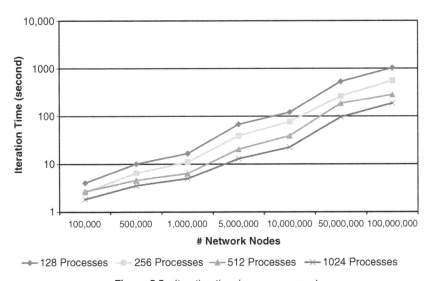

Figure 5.5 *Iteration time by process count.*

the same number of nodes run 1.62 times as fast, with maximum increases of 2.00 times and minimum of 0.90 times, still shy of perfect weak scaling. There is some very small overhead associated with data collection that increases slightly as the number of processes increases, but the variations in scaling and the lack of perfect weak scaling are most likely due to variations in the simulated network configuration. The configuration will vary between model runs

TABLE 5.1   Initialization and Run Times

Process Count	Node Count	Initialization Time	Run Time
128	100,000	1.15	4.05
128	500,000	2.52	10.11
128	1,000,000	3.25	16.73
128	5,000,000	16.15	67.86
128	10,000,000	22.59	120.81
128	50,000,000	79.40	534.08
128	100,000,000	109.08	1024.66
256	100,000	1.70	2.59
256	500,000	2.35	6.48
256	1,000,000	3.46	11.07
256	5,000,000	10.58	39.19
256	10,000,000	30.65	76.68
256	50,000,000	30.57	264.81
256	100,000,000	144.26	558.20
512	100,000	2.75	2.69
512	500,000	3.39	4.59
512	1,000,000	3.46	6.39
512	5,000,000	6.03	20.56
512	10,000,000	11.47	39.13
512	50,000,000	133.02	186.29
512	100,000,000	52.82	283.14
1024	100,000	5.15	1.85
1024	500,000	5.29	3.55
1024	1,000,000	5.70	4.98
1024	5,000,000	7.22	13.00
1024	10,000,000	8.86	22.56
1024	50,000,000	47.20	96.13
1024	100,000,000	135.33	186.36

and within the processes themselves for a single-model run. More particularly, processes whose network configuration leads to a greater number of rumored nodes take longer to complete an iteration. This, in turn, increases the idle time of other processes during schedule synchronization as they wait for the slower process to complete. We also suspect that variations in network configuration are in part responsible for the lack of perfect weak scaling when the process count is constant and the number of agents is increased, for example, the anomalous spike for 512 processes when running 50 million agents. We hope to investigate the effects of network configuration further when expanding the rumor model.

Additional future work will investigate scaling beyond 1024 processes. Scaling is of course dependent on the trade-off described earlier, especially with respect to the cross-process sharing of agents, and thus partly on the model. Increasing the number of processes can further increase interprocess communication, and the optimum number of agents then changes. Ultimately though, given that many models cannot be run at all using current ABMS tools, the trade-off is not between faster and slower, but rather between running the model or not running it all.

5.8   SUMMARY AND FUTURE WORK

This chapter has presented an overview of the Repast HPC project and, more particularly, the implementation of the Repast HPC framework. The authors hope the framework can help to leverage the increasing computing capability made available on high-performance computing platforms by bringing this capability to ABMS. Throughout this chapter, we have tried to emphasize that our concern is with creating a useful and usable working system. In this, we differ from much of the related work in this area. Our experience with the implementation of previous Repast systems has taught us about the natural fit between agent-based modeling and object-oriented programming, the importance of flexible and dynamic scheduling, and, perhaps most importantly, that the implementation should be driven by modeling rather than by framework concerns. All these we have tried to carry over into the larger-scale distributed context.

In particular, the latter item informs the implementation of the RepastProcess, the distributed projections, and the scheduling implementation. The details of the parallel implementation and MPI are hidden behind a simpler API. Where parallel and distributed framework concerns are unavoidable, an effort has been made to make these more conceptually accessible by framing the documentation and API in terms of the executing process, local and nonlocal shared agents. The source code of the example rumor model application neatly reflects this. The intention of the model itself is easily discernable and the parallel or distributed parts of the code are most naturally interpreted in terms of executing process, local and nonlocal shared agents. Future work will focus on expanding the rumor model, developing additional models using the framework, and on improvements to and experiments with the framework itself. The intent is to fill existing gaps and to leverage the excellent existing work in parallel with discrete-event scheduling to improve the scheduling architecture. The simple sample rumor model has illustrated the trade-off between agent population size and interprocess communication. Future work will attempt to minimize this trade-off, allowing for scaling to a larger number of processes with equivalent-sized agent populations.

REFERENCES

R. L. Bagrodia and W. Liao. Parallel simulation of the Sharks World problem. In *Proc. of the 22nd Winter Simulation Conference*, pp. 191–198, 1990.

A. Barabási and R. Albert. Emergence of scaling in random networks. *Science*, 286(5439):412–413, 1999.

D. W. Bauer, C. D. Carothers, and A. Holder. Scalable time warp on Blue Gene supercomputers. In *2009 ACM/IEEE/SCS 23rd Workshop on Principles of Advanced and Distributed Simulation*, pp. 35–44, 2009.

J. Bradshaw. KAoS: An open agent architecture supporting reuse, interoperability, and extensibility. In *Proc. of the 1996 Knowledge Aquisition Workshop*, 1996.

L. Briesemeister, P. Lincoln, and P. Porras. Epidemic profiles and defense of scale-free networks. In *Proc. of the 2003 ACM Workshop on Rapid Malcode*, 1998.

C. D. Carothers, D. W. Bauer, and S. O. Pearce. ROSS: A high-performance low memory, modular time warp system. *Journal of Parallel and Distributed Computing*, 62(11):1648–1669, 2002.

K. M. Chandy and J. Misra. Distributed simulation: A case-study in design and verification of distributed programs. *IEEE Transactions on Software Engineering*, SE(5):440–452, 1979.

K. M. Chandy and J. Misra. Asynchronous distributed simulation via a sequence of parallel computations. *Communications of the ACM*, 24(4):198–205, 1981.

R. M. D'Souza, M. Lysenko, and K. Rahmani. SugarScape on steroids: Simulating over a million agents at interactive rates. In *Proc. of the Agent 2007 Conference on Complex Interaction and Social Emergence*, 2007.

S. Deissenberg and H. van der Hoog. EURACE: A massively parallel agent-based model of the European economy. Technical report, Universites d'Aix-Marseille, 2009. halshs.archives-ouvertes.fr/docs/00/33/97/56/PDF/ DT2008-39.pdf.

R. M. Fujimoto. Parallel discrete event simulation. *Communications of the ACM*, 33(10):30–53, 1990.

R. M. Fujimoto. *Parallel and Distributed Simulation Systems*. Piscataway, NJ: Wiley Interscience, 2000.

N. Houari and B. H. Far. Building collaborative intelligent agents: Revealing main pillars. In *Proc. of the Canadian Conference on Electrical and Computer Engineering*, 2005.

T. R. Howe, N. T. Collier, M. J. North, et al. Containing agents: Contexts, projections and agents. In *Proc. of the Agent 2006 conference on social agents: Results and prospects*, 2006.

P. Hudak. Conception, evolution, and application of functional programming languages. *ACM Computing Surveys*, 21(3):359–411, 1989.

D. R. Jefferson. Virtual time. *ACM Transactions of Programming Languages and Systems*, 7(3):404–425, 1985.

D. Jefferson, B. Beckman, F. Wieland, et al. Time warp operating system. *ACM SIGOPS Operating Systems Review*, 21(5):77–93, 1987.

V. Jha and R. Bagrodia. A unified framework for conservative and optimistic distributed simulation. *ACM SIGSIM Simulation Digest*, 24(1):12–19, 1994.

H. Karimabadi, J. Driscoll, J. Dave, et al. Parallel discrete event simulation of grid-based models: Asynchronous electromagnetic hybrid code. *Springer Verlag Lecture Notes in Computer Science*, 3732:573–582, 2006.

K. Klemm and V. Egužluz. Highly clustered scale free networks. *Physical Review E*, 65(036123):1–6, 2002.

M. T. Koehler and B. F. Tivnan. Clustered computing with NetLogo and Repast J: Beyond chewing gum and duct tape. In *Proc. of the Agent 2005 Conference on Generative Social Processes, Models, and Mechanisms*, pp. 43–54, 2005.

Y.-K. Kwok and K.-P. Chow. On load balancing for distributed multiagent computing. *IEEE Transactions on Parallel and Distributed Systems*, 13(8):787–801, 2002.

G. Laycock. The theory and practice of specification based software testing. PhD thesis, University of Sheffield, 1993.

M. Lysenko and R. M. D'Souza. A framework for meagascale agent-based model simulations on graphical processing units. *Journal of Artificial Societies and Social Simulation*, 11(4), 2008. Available at http://jasss.soc.surrey.ac.uk/11/4/10.html.

F. Massaioli, F. Castiglione, and M. Bernashci. OpenMP parallelization of agent-based models. *Parallel Computing*, 31(10–12):1066–1081, 2005.

M. Matsumoto and T. Nishimura. Mersenne Twister: A 623-dimensionality equdistributed uniform psuedo-random number generator. *ACM Transactions on Modeling and Computer Simulation: Special Issue on Uniform Random Number Generation*, 8(1):3–30, 1998.

D. M. Nicol and S. E. Riffe. A "conservative" approach to parallelizing the Sharks World simulation. In *Proc. of the 22nd Winter Simulation Conference*, 1990.

M. J. North and C. M. Macal. *Managing Business Complexity: Discovering Strategic Solutions with Agent-Based Modeling and Simulation*. New York: Oxford University Press, 2007.

M. J. North, E. Tatara, N. T. Collier, et al. Visual agent-based model development with Repast Simphony. In *Proc. of the Agent 2007 Conference on Complex Interaction and Social Emergence*, 2007.

T. Oguara, D. Chen, G. Theodoropoulos, et al. An adaptive load management mechanism for distribution simulation of multi-agent systems. In *Proc. of the 9th IEEE Int'l Symposium on Distributed Simulation and Real-Time Applications*, 2006.

K. S. Perumulla. μsik—a micro-kernel for parallel/distributed simulation systems. In *Workshop on Principles of Advanced and Distributed Simulations*, 2005.

K. S. Perumulla. Parallel and distributed simulation: Traditional techniques and recent advances. In *Proc. of the Winter Simulation Conference (WSC)*, 2006.

K. S. Perumulla. Scaling time warp-based discrete event execution to 10^4 processors on a Blue Gene supercomputer. In *Proc. of the 4th Int'l Conference on Computing Frontiers*, pp. 69–76, 2007.

M. T. Presley, P. L. Reiher, and S. F. Bellenot. A time warp implementation of the Sharks World. In *Proc. of the 22nd Winter Simulation Conference*, pp. 199–203, 1990.

M. Scheutz, P. Schermerhorn, R. Connaughaton, et al. SWAGES: An extendable distributed experimentation system for large-scale agent-based a life simulations. In *Proc. of the 10th Int'l Conference on the Simulation and Synthesis of Living Systems*, 2006.

D. Zurell, U. Berger, J. S. Cabral, et al. The virtual ecologist approach: Simulating data and observers. *Oikos—A Journal of Ecology*, 119(4):622–635, 2010.

Chapter **6**

Building and Running Collaborative Distributed Multiscale Applications

Katarzyna Rycerz

AGH University of Science and Technology, Krakow, Poland, and ACC Cyfronet AGH, Krakow, Poland

Marian Bubak

AGH University of Science and Technology, Krakow, Poland, and Institute for Informatics, University of Amsterdam, The Netherlands

6.1 INTRODUCTION

Multiscale simulations are a dynamically developing area of complex system modeling. Examples include blood flow simulations (e.g., in treatment of in-stent restenosis) as presented by Caiazzo et al. (2009), solid tumor models (Hirsch et al., 2009), or stellar system simulations (Portegies Zwart et al., 2008). Modern technical solutions provided by computer science offer many useful features for such simulations, including support for composability and reuse (component approach) as well as resource sharing (grid concept). In this chapter, we explain how to efficiently exploit some of them. Providing a distributed, easy-to-use e-infrastructures that can be collaboratively and transparently shared by multidisciplinary scientists is one of the main aspects of large-scale computing. As shown in this chapter, some of these solutions can also be successfully applied to the field of multiscale simulations.

Large-Scale Computing, First Edition. Edited by Werner Dubitzky, Krzysztof Kurowski, Bernhard Schott.
© 2012 John Wiley & Sons, Inc. Published 2012 by John Wiley & Sons, Inc.

Grid technologies can be exploited for the development and execution of multiscale simulations on various levels. The first approach, based on the concept presented by Foster et al. (2001), is to use the grid as a metacomputer with access to distributed computing and storage resources. Groen et al. (2010) presents a case where a galaxy collision simulation switches at run time between heterogeneous resources on two grid sites: a graphical processing unit (GPU)-enabled computer (Pharr and Fernando, 2005) in The Netherlands and a Gravity Pipe (GRAPE)-enabled machine (Makino et al., 2003) in the United States.

The second approach to large-scale distributed computing, based on the concept of Web[1] and grid services (Foster et al., 2002), allows the user to construct an application from software components (services) installed in the e-infrastructure. In this approach, higher-level software services are shared in addition to pure computational and storage resources. A good example is the GridSpace virtual laboratory presented in Malawski et al. (2010).

The third approach, also connected with the idea of cloud computing (Rajkumar and Rajiv, 2010), is to enable the user to dynamically deploy software/applications into high-level containers residing on the e-infrastructure. One example is the H2O framework (Kurzyniec et al, 2003), where pluglets written by a user are deployed into so-called kernels installed on the grid. Another system that can be mentioned here is the Highly Available Dynamic Deployment Infrastructure for Globus Toolkit (HAND) for dynamic deployment of grid services (Qi et al., 2007).

In this chapter, we focus on applying the second approach to support multiscale simulations, that is, allowing the user to build an application from software components or services installed in the e-infrastructure. We propose a component environment for such simulations that allows multiscale application developers to wrap their models as recombinant software components and share them using the available e-infrastructure. This would help scientists working on multiscale problems to exchange models and connect them in various applications.

This chapter is organized as follows. Section 6.2 defines the requirements of complex multiscale simulations. Section 6.3 presents some of the available grid and component technologies and some related work. Section 6.4 describes an environment supporting the high-level architecture (HLA) component model. Section 6.5 describes an experiment involving a sample multiscale simulation of a dense stellar system. Summary and future work are presented in Section 6.6.

6.2 REQUIREMENTS OF MULTISCALE SIMULATIONS

When designing an environment supporting the creation and execution of multiscale simulations, one can identify two different types of issues. One

---

[1] Web services: http://www.w3.org/TR/ws-arch/.

group of issues, described in Section 6.2.1, relates to the problem of actually connecting two or more simulation modules together. The set of actual requirements depends on the type of simulation. In this chapter, we focus on advanced time synchronization requirements and describe three different types of time interactions between multiscale simulation models. Issues related to composability and reusability of existing simulation modules are described in Section 6.2.2, where we also justify the need for wrapping simulation models into recombinant software components that can later be selected and assembled in various combinations.

6.2.1 Interactions between Single-Scale Models

Interactions between single-scale modules require adherence to conservation laws, support for joining modules with different internal time management, efficient data exchange between pairs of modules with different scales, and so on. The set of actual requirements depends on the type of simulation: Not every set of models requires advanced time management, so the simulation developer needs to make an optimal choice between functionality and simplicity of the environment. For example, the Multiscale Multiphysics Scientific Environment (MUSE) (Portegies Zwart et al., 2008) requires advanced time synchronization between modules, whereas the simulation model of in-stent restenosis (Caiazzo et al., 2009), applying asynchronous unidirectional pipes blocking on the receiver site, remains sufficient for communication.

In our work, we focus on models that require advanced synchronization support. Temporal interactions occurring between multiscale components were analyzed in our previous study (Rycerz et al., 2008c); they will also be briefly described in the following three subsections. We can distinguish scenarios where modules are running concurrently and exchanging time-stamped data using (1) a conservative approach, (2) an optimistic approach, and (3) situations where one module triggers another and waits for its data.

Our analysis bases on modules derived from MUSE, which is a sequential simulation environment designed for dense stellar systems. However, the types of supported interactions are generic and can apply to other module connections that require advanced time management facilities.

6.2.1.1 Concurrent Execution: Conservative Approach   One of the most common types of interactions between simulation models occurs when one simulation regulates the timing of another simulation, as shown in Figure 6.1 The conservative constrained simulation can move forward as far as it is allowed by the regulating simulation. In Figure 6.1, the constrained simulation is not allowed to go beyond time $t = t_1$ until it is certain that it will not receive any messages with time stamp $t = t_1$ (i.e., until the regulating simulation does not move forward).

This type of interaction is useful when data sent by the regulating simulation affect the entire constrained simulation and any rollback would be too

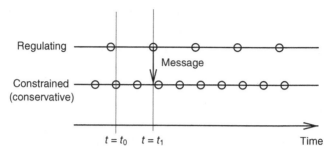

Figure 6.1 Concurrent execution—conservative approach. The conservative constrained simulation can move forward as far as it is allowed by the regulating simulation.

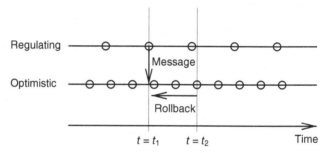

Figure 6.2 Concurrent execution—optimistic approach. Both simulations can advance in time as far as needed, but a rollback becomes necessary if an outdated message is received.

costly. In multiscale simulations, this approach can be applied when a macroscale model produces some useful data for a mesoscale model. It enables the macroscale model simulation to move forward as far as needed and to control the mesoscale model simulation. As the mesoscale model moves forward slower (in terms of simulation time units), controlling it via the macroscale model should be acceptable.

6.2.1.2 Concurrent Execution: Optimistic Approach
When a conservative approach is significant for simulation execution, one may apply optimistic model interaction. In this case, both simulations can advance in time as far as possible, but once one of them receives a message with a time stamp lower then its execution time, it needs to roll back its state to match this execution time. This is shown in Figure 6.2, where the optimistic simulation has to roll back from $t = t_2$ to $t = t_1$ because the regulating simulation has just sent a message with time stamp $t = t_1$.

This type of interaction is useful for models where rollback is relatively inexpensive—for example, an outdated message only affects some of the

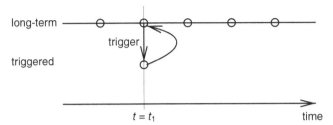

Figure 6.3 *Execution triggered by another simulation. One simulation launches the execution of another and waits for its results.*

results already produced. In multiscale simulations, this situation occurs, for instance, when the microscale model produces data useful for the macroscale model in an asynchronous way and such data only affect some part of the macroscale model.

6.2.1.3 Execution Triggered by Another Execution Sometimes, it is not necessary to run two simulation models concurrently, either in an optimistic or conservative way. Consequently, a much simpler solution can be applied. This happens when model A "knows" when to launch the execution of model B (usually with a relatively short execution time) and waits for its results, as presented in Figure 6.3, where the long-term simulation triggers a short model execution.

In multiscale simulations, this situation occurs, for example, when a mesoscale model needs to launch a microscale simulation to obtain data that will have a significant effect on the whole mesoscale simulation.

6.2.2 Interoperability, Composability, and Reuse of Simulation Models

Another important set of requirements involves the reuse of existing models and their composability. Specifically, it might prove useful to find existing models and to link them together. Therefore, there is a need for wrapping simulations into recombinant components that can be selected and assembled in various ways to satisfy specific requirements. The modeling and simulation community has a need for simulation interoperability and composability.[2]

Ongoing efforts in this research area have resulted in the Base Object Model (BOM)[3] standard for semantic description of simulation components and their relationships. However, none of the present solutions is widely applied by simulation developers, and they generally do not address issues specific to multiscale simulations. This is due to the fact that composability and

---

[2] The Simulation Interoperability Standards Organization (SISO) home page: http://www.sisostds.org/.
[3] BOM standards home page: http://www.boms.info/standards.htm.

reuse requirements imply simplicity and ease of use, which, in turn, has to be provided along with advanced functionality, specific to multiscale simulations.

The above-mentioned requirements can be fulfilled by applying new technological solutions, assisting multiscale simulation researchers in collaboration, exchange of simulation models, and sharing know-how ideas for the effective operation of such complex simulations. The next section provides an overview of new technological solutions.

6.3 AVAILABLE TECHNOLOGIES

6.3.1 Tools for Multiscale Simulation Development

6.3.1.1 Model Coupling Toolkit (MCT) MCT (Larson et al., 2005) is a tool capable of simplifying the construction of parallel coupled models. It was successfully exploited in a parallel climate model. MCT applies the Message Passing Interface (MPI) style of communication between simulation modules. It is designed for the domain data decomposition of a simulated problem and provides support for advanced data transformations between different modules. Data are exchanged between modules with a defined coupling frequency.

6.3.1.2 HLA HLA[4] is a standard for large-scale distributed interactive simulations. HLA offers the ability to plug and unplug modules containing various simulation models, with different internal types of time management, to/from a complex simulation system. The elements of the simulation system are connected together in a tuple space, used for subscribing to and publishing time-stamped events and data objects.

6.3.1.3 MUSE MUSE (Portegies Zwart et al., 2008) is a software environment for astrophysical applications in which different simulation models of star systems are incorporated into a single framework. The scripting approach (Python) is used to couple software modules containing models. Currently, the set of available models includes stellar evolution, hydrodynamics, stellar dynamics, and radiative transfer domains. Python scripts are used to execute modules in an appropriate order. Execution is sequential; that is, in a single time loop, the module containing model A is executed followed by the module with model B. Finally, the data are updated in both modules.

6.3.1.4 The Multiscale Coupling Library and Environment (MUSCLE) MUSCLE (Hegewald et al., 2008) provides a software framework for building simulations according to the complex automata theory (Hoekstra et al., 2007).

---

[4] High Level Architecture specification—IEEE 1516.

TABLE 6.1   Comparison of Environments Supporting Multiscale Model Coupling

	Type of Execution	Time Management	Data Distribution	Data Converters between Modules
MCT	Parallel	Data exchanged with defined frequency	Domain decomposition, message passing (MPI)	Support for data conversion between processes
HLA	Concurrent execution in distributed environment	Advanced time management (conservative/ optimistic)	Publish/subscribe in tuple space	No explicit support for data conversion
MUSE	Sequential, code coupled by Python	Sequential execution of model parts in a time loop	Explicit function calls that transform data from one model to another in the time loop	Data conversions can be programmed in functions
MUSCLE	Concurrent execution in distributed environment	Asynchronous (nonblocking send-blocking receive)	Unidirectional pipes (conduits) between models	Data converters can be applied to conduits

It has been applied to the simulation of coronary artery in-stent restenosis (Caiazzo et al., 2009), but it is designed for complex automata in general. The framework introduces the concept of kernels, which communicate by unidirectional pipelines dedicated to passing specific types of data from/to a kernel. Communication is asynchronous and blocks only on the receiver side.

Table 6.1 depicts data comparing the environments mentioned earlier. The data show that the functionality of the environments is complementary and that the user choice depends on the concrete simulation requirements. MCT is appropriate for parallel simulations with domain decomposition, while HLA is suitable for joining simulation with different time management polices, but is also quite complicated and involves a steep learning curve. MUSE is relatively easy to use, but it is designed mainly for astrophysical simulations. MUSCLE was designed for cellular automata problems and can be used for simulations that require simple asynchronous communication.

In our work, we have decided to apply HLA due to its advanced support for distributed simulations, especially its time management facilities, which cover all three types of time interactions mentioned in Section 6.2.1), and it also allows us to join different types of time interactions in one simulation system.

6.3.2   Support for Composability

Composition approaches vary with regard to their degree of coupling. The most loosely coupled approach includes workflow models, such as Business

Process Execution Language (BPEL),[5] ASKALON (Fahringer et al., 2005), or Kepler (Altintas et al., 2004). They focus mainly on loosely coupled services without support for direct links. Script-based solutions for loosely coupled compositions include Geodise (Pound et al., 2003) and GridSpace virtual laboratory.[6]

In a tightly coupled approach to composition, the application is presented as a graph of connected components. The graph defines the workflow of the application, showing how components directly call each other. There are also sample architectures, for example, the Service Component Architecture (SCA),[7] where business functionality is provided as a series of services, which are assembled together to create solutions that serve a particular business need. Another example is the Common Component Architecture (CCA) described by Armstrong et al. (2006) (with implementations such as MOCCA, described by Malawski et al., 2006), used in high-performance computing, where scientific components are directly connected by their *Uses* and *Provides* ports. The Provides port is a set of public interfaces that the component implements and that can be referenced and "used" by other components. The Uses port can be viewed as a connection point on the surface of the component where the framework can attach (connect) references to Provides ports implemented by other components of the framework.

In addition, there are also high-level composition models based on a skeleton concept, such as those recently developed for grid computing (e.g., grid component model [GCM]), presented by Françoise et al., 2009). To the best of our knowledge, none of the existing component approaches provides advanced features for distributed multiscale simulations (in particular, they do not explicitly support connections of modules with different time management mechanisms); thus, there is a need for a component model that supports such advanced, simulation-specific types of connections.

6.3.3 Support for Simulation Sharing

The grid concept (Foster et al., 2002) is oriented toward joining geographically distributed communities of scientists working on shared problems. This feature also enables users of distributed simulations to more easily exchange software components containing their models. The grid aims at facilitating access to distributed resources across administrative domains. For example, the concept of a virtual organization (VO) introduced by Foster et al. (2001) allows a set of individuals and/or interrelated institutions to share resources, services, and applications regardless of their administrative domains and geographic locations.

[5] OASIS WSBPEL Technical Committee. Web Services Business Process Execution Language Version 2.0, April 2007. http://docs.oasis-open.org/wsbpel/2.0/OS/wsbpel-v2.0-OS.html.
[6] Virolab home page: http://virolab.cyfronet.pl.
[7] G. Barber. Service Component Architecture Home, 2007. http://osoa.org/display/Main/Service+Component+Architecture+Home.

One of the most important implementations of the grid concept is the Globus Toolkit.[8] To achieve interoperability between distributed systems, the Globus community has come up with the idea of extending the Web services standard[9] with state and dynamic creation facilities, which resulted in the Web Services Resource Framework (WSRF) standard. WSRF uses a classical Web service as an interface to a stateful WS-Resource.[10] Apart from the work of the Globus community, there are also other approaches to grid computing. Grids can be built on top of the H2O resource sharing platform (Kurzyniec et al., 2003), which introduces the concept of kernels, that is, containers for user-provided software modules. Usually, kernels should be installed on an e-infrastructure, while the modules (called pluglets) can be dynamically deployed by authorized third parties (not necessarily kernel owners). The framework is portable (Java based), secure, scalable, and lightweight. Finally, grid architectures can be constructed from pure Web services or component technologies described in the previous sections.

Examples of working grids include European e-infrastructures such as Enabling Grids for E-sciencE (EGEE) or Distributed European Infrastructure for Supercomputing Applications (DEISA).

As our goal is to support component exchange between scientists working in the field of multiscale simulations regardless of their actual geographic location, we have decided to use grid solutions and have chosen the H2O framework as a lightweight platform designed to support composition frameworks.

6.4 AN ENVIRONMENT SUPPORTING THE HLA COMPONENT MODEL

6.4.1 Architecture of the CompoHLA Environment

The architecture of the CompoHLA environment is shown in Figure 6.4. A *Client CompoHLA Manager* provides a user API accessible via a scripting language to call other modules of the framework. The *Remote CompoHLA Manager* manages simulation configuration files—currently, an HLA federation object model (FOM) that describes events and objects exchanged between simulation modules. It also starts and stops a legacy run-time infrastructure (RTI) control process necessary to run an HLA application.

HLA components contain actual simulation models wrapped as software modules, plugged into the HLA RTI by join/resign mechanisms, which they use to communicate. In comparison, the CCA model introduces direct connections between components by joining Provides and Uses ports, as described in Section 6.3.2. The components can use all available mechanisms provided

---

[8] http://www.globus.org.
[9] Web services: http://www.w3.org/TR/ws-arch/.
[10] WSRF home page: http://www.globus.org/wsrf/.

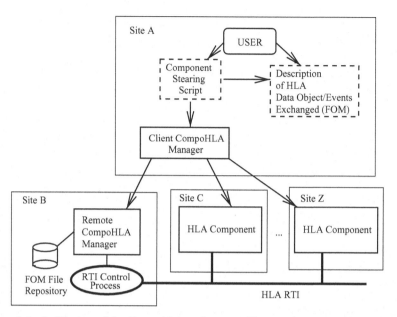

Figure 6.4 *Architecture of the CompoHLA environment. The user interacts with HLA components and the RTI control process via the Client CompoHLA Manager API accessible via a scripting language.*

by HLA (time and data management), especially when building multiscale simulations with different timescales. Thanks to HLA features it is possible to aggregate models with different internal timescales into one coherent simulation system. Different types of time management (event driven vs. time driven and conservative vs. optimistic) can be integrated so that all time interactions described in Section 6.2.1 are supported. Simulation modules can exchange events and data objects using publish/subscribe mechanisms. For advanced synchronization, they can use time-stamped messages (transitory events and persistent objects).

The difference between this approach and the original HLA is that the behavior of components and their interactions is defined and set by an external module upon user request. The proposed approach separates component developers from users who wish to construct a particular distributed simulation system from existing components. Preliminary component design and implementation are described in our previous work (Rycerz et al., 2008a, 2008b).

6.4.2 Interactions within the CompoHLA Environment

A generic time sequence diagram showing an interaction scenario between components in the CompoHLA environment during the setup process is pre-

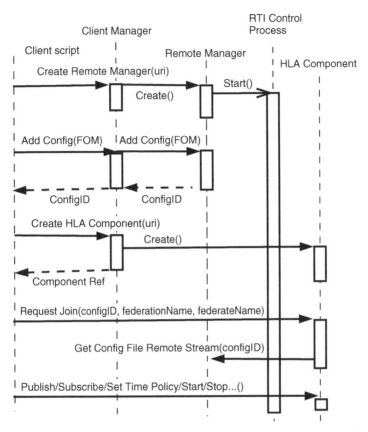

Figure 6.5 *Time sequence diagram showing an interaction scenario within the CompoHLA environment when setting up a simulation from HLA components.*

sented in Figure 6.5. A client script calls the Client CompoHLA Manager to create the Remote CompoHLA Manager, which, in turn, starts the legacy RTI control process. Subsequently, once the user prepares a FOM file describing data objects and events to be exchanged between *HLA components* taking part in the simulation, the client script asks the Client CompoHLA Manager to add the appropriate FOM as a config file. The Remote CompoHLA Manager, which manages all config files, assigns an appropriate ID and returns it. When this is done, HLA components can be created and asked to join. Each join request includes the ID of the config file, which is then fetched by the HLA component from the Remote CompoHLA Manager. Given this file, the actual join operation for the legacy RTI control process can be performed later by an internal component scheduler as descibed in Section 6.4.3. Finally, other operations can be invoked on the component (publish, subscribe, set time policy, start, stop, etc.)

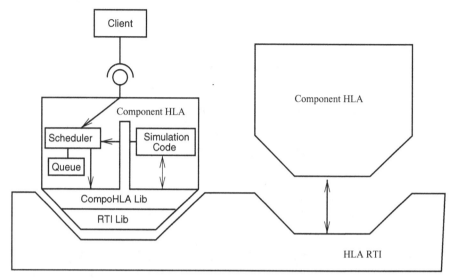

Figure 6.6 *Architecture of the HLA component plugged into the HLA RTI communication bus. The component exposes an interface for external requests. Thus, it is possible to change the state of the simulation in the HLA RTI layer by using an independent client.*

6.4.3 HLA Components

The detailed architecture of an HLA component is shown in Figure 6.6. The simulation code (custom for each component and provided by the component developer) interfaces other elements of the component by means of the CompoHLA library.

Components are plugged into the HLA RTI to communicate with each other by means of HLA mechanisms. They need to process internal requests coming from the simulation code layer and the HLA RTI layer (from other components). Moreover, they expose interfaces for external state change requests in the HLA RTI layer generated from the client while the actual simulation is already running.

To efficiently handle concurrent processing of internal and external requests, we have applied ideas from the active object pattern presented by Lavender and Schmidt (1995), which separates invocation from execution. We have introduced a scheduler processing external requests, called from the simulation by a single routine (Pycerz and Bubak, in print). External requests are processed asynchronously—the scheduler places them in a queue, where they can later be retrieved and actually executed (when called from the simulation code).

The scenario of interaction between HLA component modules is shown in Figure 6.7 in the form of a time sequence diagram.

First, the client invokes a request on the HLA component that is processed by the ComponentHLA class. The ComponentHLA class invokes the enqueue

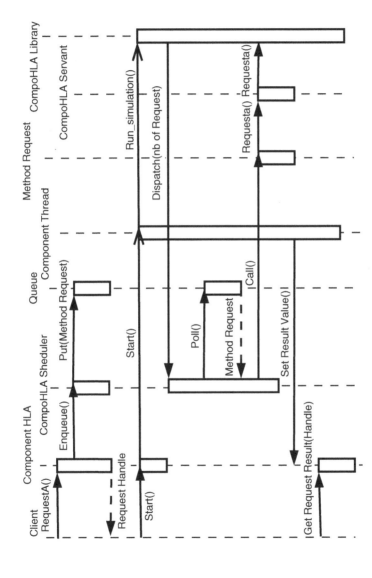

Figure 6.7 Time sequence diagram showing the interaction scenario between HLA component modules compliant with the active object pattern (Lavender and Schmidt, 1995). External requests are placed in a queue. Once the simulation starts, the scheduler is asked to dispatch the requests.

method of `CompoHLAScheduler`, which places the appropriate type of request (`MethodRequest`) in the queue. The handle of this request is returned to the client. If the client invokes the `start` method on the HLA component, it causes `ComponentHLA` to start a new thread, `ComponentThread`, which runs the actual simulation (using the CompoHLA library API). During the execution of the simulation, the user code (via CompoHLA library API) periodically calls the `dispatch()` method of the `CompoHLAScheduler`. The scheduler withdraws `MethodRequests` from the queue and executes their `call` method, which then executes appropriate request methods of `CompoHLAServant` (according to the actual type of `MethodRequest`). The server calls the CompoHLA library, which actually executes the request. Results are stored in `ComponentHLA` by calling its `setResultValue` method, which enables them to be retrieved by the client.

6.4.4 CompoHLA Component Users

There are two different group of users of the CompoHLA system: developers who create components and scientists who actually use them (component users).

The component developer can wrap a custom legacy simulation module using the CompoHLA library as described by Rycerz et al. (2008a, 2008b) so that it can be easily plugged into the distributed multiscale simulation. The use of the compoHLA library does not free the developer from understanding HLA time management and data exchange mechanisms; however, it considerably simplifies their use and enables the HLA component to be steered from the user code.

The component user can use a scripting language to dynamically set up a distributed multiscale simulation comprising selected HLA components. He can also decide which components will participate in simulation and can dynamically plug and unplug them to/from running simulations. Additionally, the user can decide how components will interact with one another (e.g., by setting up appropriate HLA subscription/publication and time management mechanisms) and can change the nature of their interactions during run time.

6.5 CASE STUDY WITH THE MUSE APPLICATION

The implementation of the CompoHLA environment is based on the H2O technology. The functionality of the CompoHLA system is presented here on the example of a multiscale dense stellar system simulation. We have used simulation modules with different timescales taken from MUSE, as described in Section 6.3.1. Our experiment compares the execution of legacy sequential MUSE with its distributed version using HLA components on the grid. We make use of the Dutch Grid DAS3[11] infrastructure. Components are built from

---

[11] The Distributed ASCI Supercomputer 3 Web site: http://www.cs.vu.nl/das3.

two MUSE modules: star evolution simulation (macroscale) and star dynamics simulation (mesoscale), which run concurrently to show the second type of time interaction described in Section 6.2.1.1, namely, concurrent execution—conservative approach. This experiment is a good example of using the HLA time management features when one simulation controls the timing of another. The dynamics module should obtain updates from the evolution module before it actually passes the appropriate point in time. The HLA time management mechanism (High Level Architecture specification—IEEE 1516) of the *regulating* federate (evolution), which controls the time flow in the *constrained* federate (dynamics), is therefore very useful. We have created a sample simulation consisting of a hundred stars. Data representing mass, radius, 3-D position, and velocity of stars were exchanged between the evolution and dynamics modules. The HLA components were asked to join the simulation (both), set the time-regulating policy, and publish data (evolution component) or subscribe to data (dynamics component). Subsequently, the components were asked to unset their time policy, resign the simulation, and stop. Figure 6.8 shows a sample user script (JRuby file) that performs these steps.

In our implementation, we have used H2O v2.1 and HLA CERTI implementation v3.3.2. The client script was written in JRuby 1.5.0.RC3 (Ruby 1.8.7). Experiments were performed using DAS3. MUSE sequential execution was run on a grid node in Delft. The actual setup of the HLA component experiment is shown in Figure 6.9.

The component client was run on a grid node in Leiden; the dynamics component was run on a grid node in Delft; the evolution component was run at UvA, Amsterdam, while the HLA control process was run at Vrije University, Amsterdam. All grid notes shared a similar architecture (dual AMD Opteron compute nodes, 2.4 GHz, 4-GB RAM). DAS3 employs a novel internal wide-area interconnect based on light paths between its grid sites.[12] A comparison between the distributed version using HLA components and the sequential

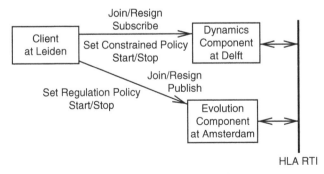

Figure 6.8 *A fragment of a sample user JRuby script. The script show the setting up and steering of a multiscale simulation consisting of two HLA components that contain dynamics (mesoscale) and evolution (macroscale) models.*

[12] StarPlane home page http://www.starplane.org/.

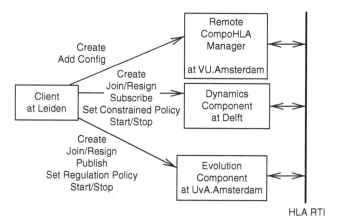

Figure 6.9 *Setup of the HLA component experiment. Elements of the CompoHLA environment are installed on different sites on the DAS3 e-infrastructure and are called by the client script.*

TABLE 6.2   Stellar Dynamics and Evolution: Multiscale Simulation Time Using Original MUSE Sequential Execution Steered by Python Scheduler

Sequential MUSE

	Time (second)	σ
Dynamics calculation	18.1	0.6
Evolution calculation	0.004	0.0001
Data update evolution–dynamics	0.01	0.001
Total simulation time	18.1	0.6

TABLE 6.3   Stellar Dynamics and Evolution: Multiscale Simulation Time Using Distributed HLA Components

Dynamics Component			Evolution Component		
	Time (second)	σ		Time (second)	σ
Dynamics calculation	17.6	0.08	Evolution calculation	0.004	0.0001
Synchronization with evolution	0.007	0.0001	Synchronization with dynamics	0.04	0.001
Scheduler	0.6	0.05	Scheduler	0.6	0.05
Total simulation time	18.3	0.08	Total simulation time	1.1	0.05

MUSE Python scheduler is shown in Tables 6.2 and 6.3. The overhead of using the scheduler and HLA communication mechanisms (time management and data transfer) was very low in comparison to actual calculations. Note that the computation time for the dynamics component was far more significant than evolution time; therefore, the dynamics were the dominant factor of the total simulation time in both cases (concurrent and sequential execution). The per-

TABLE 6.4   Timing of Requests to the Modules of the
CompoHLA Environment Issued from the JRuby Script

Request	Time (millisecond)	σ
Create remote manager	2020	36
Create dynamics component	2120	23
Create evolution component	2150	27
Add config to the manager	150	39
Join evolution	140	9
Join dynamics	171	3
Publish evolution	4	0.1
Subscribe dynamics	5	0.4
Set time policy evolution	4	0.1
Set time policy dynamics	5	0.4
Start evolution	4	0.4
Start dynamics	7	0.6
Unset time policy evolution	5	0.4
Unset time policy dynamics	9	0.4
Resign evolution	5	0.2
Resign dynamics	8	0.9
Stop evolution	3	0.4
Stop dynamics	8	0.4

formance results prove that the overhead of the HLA-based distribution (especially its repeating part involving synchronization between multiscale elements) was low and that HLA can be beneficial for multiscale simulations. Moreover, the scheduler-based deign, which manages external requests, did not produce much overhead.

Table 6.4 shows the execution time for requests issued to the HLA component from the client perspective, measured in a JRuby script from Figure 6.8 (in miliseconds). The most time-consuming part was the creation of a remote manager and components as it required a loading code. Adding a config file to the manager and preparing join requests was also time-consuming as these operations required transferring config files.

Other requests to the components were on the order of several milliseconds. The requests were called in an asynchronous fashion so that during the call, they were only scheduled in the queue. Actual work was performed by the scheduler, with timing presented in Table 6.3. The results of the experiment showed that the execution time of requests was relatively low and that the component layer did not introduce significant overhead.

6.6   SUMMARY AND FUTURE WORK

In this chapter, we presented an environment supporting the HLA component model that enables the user to dynamically compose/decompose distributed simulations from multiscale elements residing on e-infrastructures. The simulation models, encapsulated in software components, can make use of advanced HLA mechanisms such as time and data management, which is especially helpful when joining different timescales. The elements of the environment are programmatically accessible via scripting interfaces. Prototype functionality was described on the basis of a sample multiscale simulation of a dense stellar system in the MUSE environment. The results of the experiment show that that the component layer does not introduce much overhead.

The HLA components described in this chapter are designed to facilitate the composability of simulation models by means of HLA mechanisms, accessible and steerable from the user layer. However, in order to fully exploit their composability, we are currently in the process of integration with the GridSpace Virtual Laboratory.[13] GridSpace provides an experiment workbench for constructing experiment plans from code snippets. It supports multiple interpreters and enables access to computing infrastructures. The basic integration step is straightforward—the client script setting up HLA components (shown in Fig. 6.8) can be directly run as a GridSpace experiment and connects to the H2O component containers installed on the e-infrastructure. Integration with GridSpace yields the ability to interpret commands issued to HLA components in an interactive mode as it supports step-by-step programming where steps are not known in advance but are based on previous results. This feature is especially useful for long-running simulation models producing partial output.

We plan to use GridSpace to access other European e-infrastructures (e.g., EGEE) apart from using DAS2 and also to use the common GridSpace services (gems) for component sharing and future reuse. In the future, partial results of long-running components will be accessible directly from the GridSpace workbench using its interface to a Web application displaying output and error streams from components.

Future work will also involve a description language for connecting HLA components. Currently in our work, we use the HLA FOM definition of structures of data objects and events that need to be passed between HLA components. This approach should be further extended to provide more information related to the scale of modules. Moreover, module descriptions should be flexible enough to support simpler types of data (e.g., arrays) frequently used in legacy implementations of simulation models.

ACKNOWLEDGMENTS

The authors wish to thank Alfons Hoekstra and Jan Hegewald for discussions on MUSCLE, Simon Portegies Zwart for valuable discussions on MUSE and our colleagues for input concerning GridSpace. The authors are also very grateful to Piotr Nowakowski for his suggestions. The research presented in this chapter was partially supported by the European Union in the EFS PO KL Pr. IV Activity 4.1 Subactivity 4.1.1 project UDA-POKL.04.01.01-00-367/08-00 "Improvement of the Didactic Potential of the AGH University of Science and Technology—Human Assets" and also by the MAPPER project—grant agreement no. 261507, 7FP UE.

REFERENCES

I. Altintas, E. Jaeger, K. Lin, et al. A Web service composition and deployment framework for scientific workflows. In *ICWS'04: Proc. of the IEEE Int'l Conference on Web Services*, p. 814, Washington, DC: IEEE Computer Society, 2004.

R. Armstrong, G. Kumfert, L. Curfman McInnes, et al. The CCA component model for high-performance scientific computing. *Concurrency and Computation-Practice & Experience*, 180(2):215–229, 2006.

A. Caiazzo, D. Evans, J.-L. Falcone, et al. Towards a complex automata multiscale model of in-stent restenosis. In G. Allen et al., editors, *ICCS'09: Proc. of the 9th Int'l Conference on Computational Science*, pp. 705–714, Berlin: Springer-Verlag, 2009.

T. Fahringer, A. Jugravu, S. Pllana, et al. ASKALON: A tool set for cluster and grid computing: Research articles. *Concurrency and Computation-Practice & Experience*, 17(2–4):143–169, 2005.

I. Foster, C. Kesselman, and S. Tuecke. The anatomy of the grid: Enabling scalable virtual organizations. *International Journal of High Performance Computing Applications*, 15(3):200–222, 2001.

I. Foster, C. Kesselman, J. M. Nick, et al. The physiology of the grid: An Open Grid Services Architecture for distributed systems integration. In *Open Grid Service Infrastructure WG, Global Grid Forum*, June 2002.

B. Françoise, C. Denis, D. Cédric, et al. GCM: A grid extension to fractal for autonomous distributed components. *Annales des Télécommunications*, 64(1–2):5–24, 2009.

D. Groen, S. Harfst, and S. Portegies Zwart. The living application: A self-organizing system for complex grid tasks. *International Journal of High Performance Computing Applications*, 24(2):185–193, 2010.

J. Hegewald, M. Krafczyk, J. Tölke, et al. An agent-based coupling platform for complex automata. In M. Bubak et al., editors, *ICCS'08: Proc. of the 8th Int'l Conference on Computational Science, Part II*, pp. 227–233, Berlin: Springer-Verlag, 2008.

S. Hirsch, D. Szczerba, B. Lloyd, et al. A mechano-chemical model of a solid tumor for therapy outcome predictions. In G. Allen et al., editors, *ICCS'09: Proc. of the 9th Int'l Conference on Computational Science*, pp. 715–724, Berlin: Springer-Verlag, 2009.

A. G. Hoekstra, E. Lorenz, J.-L. Falcone, et al. Toward a complex automata formalism for multi-scale modeling. *International Journal for Multiscale Computational Engineering*, 5(6):491–502, 2007.

D. Kurzyniec et al. Towards self-organizing distributed computing frameworks: The H2O approach. *Parallel Processing Letters*, 13(2):273–290, 2003.

J. Larson, R. Jacob, and E. Ong. The Model Coupling Toolkit: A new Fortran90 toolkit for building multiphysics parallel coupled models. *International Journal of High Performance Computing Applications*, 19(3):277–292, 2005.

G. Lavender and D. C. Schmidt. Active object: An object behavioral pattern for concurrent programming. In *Proc. Pattern Languages of Programs*, 1995.

J. Makino, E. Kokubo, and T. Fukushige. Performance evaluation and tuning of GRAPE-6—towards 40 "real" Tflops. In *SC'03: Proc. of the 2003 ACM/IEEE Conference on Supercomputing*, p. 2, Washington, DC: IEEE Computer Society, 2003.

M. Malawski, T. Bartyński, and M. Bubak. Invocation of operations from script-based grid applications. *Future Generation Computer Systems*, 26(1):138–146, 2010.

M. Malawski, M. Bubak, M. Placek, et al. Experiments with distributed component computing across grid boundaries. In *Proc. of the HPC-GECO/CompFrame Workshop in Conjunction with HPDC 2006*, Paris, France, 2006.

M. Pharr and R. Fernando. *GPU Gems 2: Programming Techniques for High-Performance Graphics and General-Purpose Computation (GPU Gems)*. Boston: Addison-Wesley Professional, 2005.

S. Portegies Zwart, S. Mcmillan, B. Ó. Nualláin, et al. A multiphysics and multiscale software environment for modeling astrophysical systems. In M. Bubak et al., editors, *ICCS'08: Proc. of the 8th Int'l Conference on Computational Science, Part II*, pp. 207–216, Berlin: Springer-Verlag, 2008.

G. E. Pound, M. H. Eres, J. L. Wason, et al. A grid-enabled problem solving environment (PSE) for design optimisation within Matlab. In *IPDPS'03: Proc. of the 17th Int'l Symposium on Parallel and Distributed Processing*, p. 50.1, Washington, DC: IEEE Computer Society, 2003.

L. Qi, H. Jin, I. Foster, et al. HAND: Highly Available Dynamic Deployment Infrastructure For Globus Toolkit 4. In P. D'Ambra and M. R. Guarracino, editors, *PDP'07: Proc. of the 15th Euromicro Int'l Conference on Parallel, Distributed and Network-Based Processing*, pp. 155–162, Washington, DC: IEEE Computer Society, 2007.

B. Rajkumar and R. Rajiv. Special section: Federated resource management in grid and cloud computing systems. *Future Generation Computer Systems*, 26:1189–1191, 2010.

K. Rycerz, M. Bubak, and P. M. Sloot. HLA component-based environment for distributed multiscale simulations. In T Priol and M Vanneschi, editors, *From Grids to Service and Pervasive Computing*, pp. 229–239, Berlin: Springer-Verlag, 2008a.

K. Rycerz, M. Bubak, and P. M. Sloot. Dynamic interactions in HLA component model for multiscale simulations. In M. Bubak et al., editors, *ICCS'08: Proc. of the 8th Int'l Conference on Computational Science, Part II*, pp. 217–226, Berlin: Springer-Verlag, 2008b.

K. Rycerz, M. Bubak, and P. M. A. Sloot. Using HLA and grid for distributed multiscale simulations. In R. Wyrzykowski et al., editors, *PPAM'07: Proc. of the 7th Int'l Conference on Parallel Processing and Applied Mathematics*, pp. 780–787, Berlin: Springer-Verlag, 2008c.

Chapter 7

Large-Scale Data-Intensive Computing

Mark Parsons

EPCC, The University of Edinburgh, Edinburgh, United Kingdom

7.1 DIGITAL DATA: CHALLENGE AND OPPORTUNITY

7.1.1 The Challenge

When historians come to write the history of the early part of the 21st century, they will almost certainly describe this period as one of the enormous transformations in the way human beings interact with each other and the world around them and how they store the information arising from those interactions.

The scientific community has managed large digital data sets, in particular in the particle physics and astronomy domains, for more than 30 years. Once the preserve of the scientific domain, over the past two decades, we have witnessed the steady digitization of information throughout our daily lives. From digital photography to bar coding in shops to our electronic tax return to mobile phones, we are surrounded by data derived from digital devices. This revolution has happened surprisingly quickly and is almost certainly still in its infancy. Today, we are generating more stored data in each year than in all of the preceding years combined. We are already struggling to deal with this data deluge and, as data volumes continue to double, this problem can only get more challenging. The challenges we face include

Large-Scale Computing, First Edition. Edited by Werner Dubitzky, Krzysztof Kurowski, Bernhard Schott.
© 2012 John Wiley & Sons, Inc. Published 2012 by John Wiley & Sons, Inc.

- managing the increasing volume and complexity of primary data and the increasing rate of data derived from that primary data,
- coping with the rapid growth in users of that data who want to make use of it in their professional and private lives, and
- balancing the pressures to accommodate new requirements within existing computational infrastructures to protect existing investments.

In order to meet these challenges, a new domain of large-scale data-intensive computing is emerging where distributed data and computing resources are harnessed to allow users to manage, process, and explore individual data sets and, where appropriate, join them with other data sets to answer previously unanswerable questions.

7.1.2 The Opportunity

The massive proliferation of digital data provides mankind with previously unimaginable opportunities to understand the world around us. Those who are prepared to move adroitly in this fast-changing field will reap the greatest rewards. The key to succeeding is the acceptance that we are reaching a tipping point with regard to digital data.

Until now, for most people, the way we deal with all of the new data in our day-to-day lives both at work and at home has been to self-manage much of it. We all spend many hours archiving e-mails and photographs and moving our scientific or business data around to where we need it. As this task has grown, it has become clear to many people that self-management is no longer tenable and new tools are needed to help. However, many of these tools do not exist or are in their infancy.

There are enormous opportunities to develop new methodologies for managing and using digital data to answer a myriad of questions. These opportunities will be enabled partly through computational technologies but also through a revision of how we think about ownership, provenance, and use of data: who owns it, who created it, who can use it, and what it can be used for.

The area of data-intensive computing is a very large one, and in the remainder of this chapter, it is only possible to consider two aspects of it. In the following section, we discuss efforts to build distributed computers specifically designed to process large quantities of data. In the final section, we discussed the work of the Advanced Data Mining and Integration Research for Europe (ADMIRE) project, which is seeking to develop software tools and techniques for advanced distributed data mining.

7.2 DATA-INTENSIVE COMPUTERS

Over the past two decades, the use of massively parallel computer architectures as the basis for all supercomputing systems has become dominant. This is true both for capability computing problems (such as weather modeling), which tend to be run on tightly coupled parallel systems (traditional supercomputers), and also for capacity computing problems (such as particle physics event processing), which tend to be run on highly distributed parallel systems (grid computing systems).

Neither of these types of system are particularly good at processing large quantities of coupled data that may be stored in databases. The traditional supercomputers are largely optimized for numerical calculations and lack input/output (I/O) bandwidth from disk to processor. In these systems, data stored on disk are often stored on a coupled storage area network (SAN) system, which is not designed for rapid random access reads and writes during computation. The grid computing systems often have local disks but are connected together over low-cost networking components and, while very good at processing uncoupled data, display poor performance when individual systems have to communicate between each other over any distance. For this reason, over the past decade, a third type of massively parallel architecture has been mooted that is specifically designed to explore large amounts of scientific and other data. Such systems are called data-intensive computers.

During the 1990s, the concept of Beowulf clusters became popular. Such clusters were built from cheap commodity components (often simply racks of cheap PCs) and were designed to fill the gap between desktop PCs and much more expensive supercomputers. Jim Gray, a Microsoft Research scientist, now sadly deceased, took the original Beowulf concept and, working with Prof. Alex Szalay of the Johns Hopkins University and colleagues, created the concept of the GrayWulf data-intensive computer. This system was specifically designed for large-scale data-intensive scientific computing using databases to store the scientific data and is reported by Szalay et al. (2010). The University of Edinburgh is currently building a system that takes this idea further and explores the most effective balance of storage technologies for low-power scientific data processing.

Gene Amdahl, famous for his law describing the relationship between parallelized and sequential parts of a software application (Amdahl, 2007), also published three further measures related to the design of a well-balanced computer system. These are the following:

- **Amdahl Number:** the ratio of the number of bits of sequential I/O per second capable of being delivered to the processor divided by the number of instructions per second
- **Amdahl Memory Ratio:** the ratio of bytes of memory to instructions per second
- **Amdahl IOPS Ratio:** the number of I/O operations capable of being performed per 50,000 processor instructions

In an ideal well-balanced computer system, the Amdahl number, memory ratio, and I/O operations per second (IOPS) ratio will be close to one. In most modern systems, this is not the case. For some classes of application, this does not matter. For example, many supercomputing applications have very low ($O(10^{-5})$) Amdahl numbers because they are focused on processing numerical data held in cache memory. However, for large data sets held on disk, perhaps in databases, an Amdahl number close to one is vital. Data-intensive computer designs attempt to optimize the Amdahl number by marrying low-power processors with large amounts of fast disk (including solid state disk). This not only makes better use of the available procession power but also vastly reduces energy consumption—a key design factor as data volumes increase worldwide.

However, well-designed hardware systems are only part of the data-intensive computing story. In order to manipulate and analyze the data on such systems, it is vital to employ advanced software tools, and leading European research in this area is discussed in the next section.

7.3 ADVANCED SOFTWARE TOOLS AND TECHNIQUES

7.3.1 Data Mining and Data Integration

Using computers to search digital data for patterns leading to new information embedded in whatever the data represent has become a widely used methodology. Generally referred to as data mining, over the past 20 years, the techniques associated with it have become commonplace in both the scientific and business domains. However, most data mining is performed by bringing data together in data warehouses of some description and then mining it for information. In reality, we live in a world of highly distributed data, and integrating data in data warehouses is time-consuming and difficult to automate.

Over the past decade, a key research theme at The University of Edinburgh has been the creation of a framework for the management and integration of highly distributed data that may be stored in databases or flat files. This framework, called the Open Grid Services Architecture–Data Access and Integration (OGSA-DAI) framework,[1] has been designed to simplify the use of highly distributed data by scientific and business applications. The framework allows highly distributed data sets to be accessed over the Internet and the data integrated for analysis. The framework supports distributed database query processing and has many features that support optimization of such queries. OGSA-DAI allows applications to create *virtual* data warehouses. This is an important feature as it simplifies many of the social issues, such as ownership and access control, associated with data mining and the federation of data from multiple locations.

Although OGSA-DAI facilitates the federation of distributed data, it does little to support the analysis of that data. As discussed in Section 7.1.1, we are witnessing the creation of unprecedented amounts of digital data. In order to

---

[1] The OGSA-DAI software and detailed information are available at http://www.ogsa-dai.org.uk.

successfully use that data, we need automated methods of analysis that allow us to mine that data for information. The ADMIRE Project[2] has been designed to explore advanced scalable distributed computing technologies for data mining. The remainder of this chapter discusses the ADMIRE data mining framework, which is built on top of OGSA-DAI.

7.3.2 Making Data Mining Easier

The premise behind ADMIRE is to make distributed data mining easier for those who use it and for those who have to provide the data services they make use of. As such, the project has carefully thought about the makeup of the groups of people who engage in data mining projects. A detailed description of the ADMIRE architecture can be found in the literature (Hume et al., 2009). Each data mining community can be partitioned into three groups of experts:

- **Domain Experts.** These experts work in the particular domain where the data mining and integration (DMI) activities are to take place. They pose the questions that they want to see their data answer, question such as the following: "How can I increase sales in this retail sector?" "What are the correlations in these gene expression patterns?" or "Which geographic localities will suffer most in times of flooding?" They are presented with DMI tools optimized and tailored to their domain by the following two groups of experts. Many of the technical and operational issues are hidden from them.
- **DMI Experts.** These experts understand the algorithms and methods that are used in DMI applications. They may specialize in supporting a particular application domain. They will be presented with a tool set and a workbench, which they will use to develop new methods or to refactor, compose, and tune existing methods for their chosen domain experts (or class of domain experts). DMI experts are core users of DISPEL (the ADMIRE language) and use this to develop new methods that can be installed in the ADMIRE context. They are aware of the structures, representations, and type systems of the data they manipulate.
- **Data-Intensive Distributed Computing (DIDC) Engineers.** These experts focus on engineering the implementation of the ADMIRE platform. Their work is concerned with engineering the software that supports DMI enactment, resource management, DMI system administration, and language implementations, and which delivers interfaces for the tools that the DMI and domain experts will use. A DIDC engineer's role includes the dynamic deployment, configuration, and optimization of the data's movement, storage, and processing as automatically as possible. DIDC engineers also organize the library of data handling and data processing

---

[2] Detailed information on ADMIRE is available at http://www.admire-project.eu.

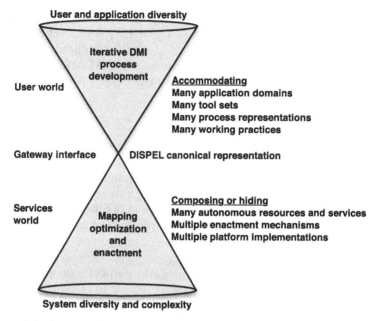

Figure 7.1 Conceptual view of ADMIRE—separating tools and enactment complexity.

activities, providing them with an appropriate computational context and methods of description.

In the same way that the people involved in DMI can be partitioned, so the ADMIRE architecture also partitions computational issues. This is crucial to simplifying and optimizing how end users can be supplied with effective DMI experiences. Computation in ADMIRE is partitioned into three levels of abstraction as a DMI process moves from creation or refinement via tools through a canonical intermediate model to a detailed representation suitable for enactment as shown in Figure 7.1:

- **Tools Level.** At this level, data sources, their components, and DMI process elements are nameable elements that users will have some knowledge of their parameters and what they do. Domain experts will use tools to compose and steer analyses such that they may be honed into tools regularly used by users. The available tools are used to study and interconnect prepackaged DMI process templates which will in general be tailored to their needs. Domain users can use the tools to specify what data resources are to be used, provide parameters, request enactments, and observe their progress. For DMI experts, the tools provide details of the representations, format, and semantics of the various data sources such that they can use the tools to specify generic patterns and to design, debug, and tune DMI processes using those patterns.

- **Canonical Intermediate Model Level.** This intermediate level is used to provide textual representations of DMI process specifications. These can be generated by the tools or by a DMI expert or DIDC engineer. These intermediate representations, in ADMIRE written in a new language called *DISPEL*, which the project has created, are then sent to an ADMIRE Gateway as a request for enactment. The gateway model is fundamental to ADMIRE because it decouples the complexity and diverse user requirements at the tools level from the complexity of the enactment level. It does this through the use of a single canonical domain of discourse represented by the DISPEL language.

- **Enactment Level.** The enactment level is the domain of the DIDC engineer. It deals with all of the software and process engineering necessary to support the full range of DMI enactments allowed by the DISPEL language and mediated through an ADMIRE Gateway. In ADMIRE, much of this level has been supported by the use of OGSA-DAI. By insulating the domain and DMI experts from this level, DIDC engineers can focus on delivering high-quality solutions to potentially complex issues while ensuring they are generic solutions that can be reused by many DMI domains.

This section is entitled "Making Data Mining Easier," which is the motto of the ADMIRE project. It would be wrong to interpret this motto as meaning it is simple to do data mining using these technologies. What ADMIRE delivers is easy-to-use patterns and models to express highly complex data mining, which would otherwise be extremely difficult to perform once and impossible to perform repeatedly.

7.3.3 The ADMIRE Workbench

The ADMIRE Workbench is central to user engagement with ADMIRE. It is described in detail by Krause et al. (2010). Figure 7.2 shows how the various workbench components are arranged and how they relate to an ADMIRE Gateway and the enactment level below this.

The components of the workbench are used to create complex data mining analyses by the domain experts and DMI experts. Much of the software engineering work to provide these tools is undertaken by DIDC engineers.

Workbench tools are used by domain or DMI experts to explore and analyze data (the *Chart Visualiser*), construct complex DMI workflows visually and edit documents written in the DISPEL language (the *Process Designer*), inspect and use context-related semantic information (the *Semantic Knowledge Sharing Assistant* [SKSA]), submit workflows to an ADMIRE Gateway and monitor its progress (the *Gateway Process Manager* [GRM]), access and query an ADMIRE registry (the *Registry Client* and *Registry View*), and visualize the data mining results (the *DMI Models Visualiser* and Chart Visualiser).

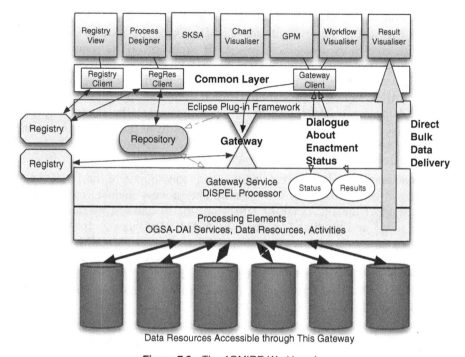

Figure 7.2   The ADMIRE Workbench.

All of the workbench tools are built as plug-ins for the well-known Eclipse software development platform. The fundamental purpose of the workbench components is to create, submit, and monitor workflow processes in the form of DISPEL documents to the Gateway Process Manager for execution. The Gateway Process Manager makes use of a common *Gateway Client* component to send the DISPEL document to the gateway and to monitor its execution. Once completed, the results can be explored in the visualizers. The purpose of the registry components is to allow useful code snippets, intermediate results, and other useful information to be stored.

The layer below the gateway represents the enactment components of the architecture. An ADMIRE Gateway processes the DISPEL documents it is presented with and produces corresponding OGSA-DAI workflows which are executed on sites and resources known to the gateway. For performance reasons, when ready, the results are delivered directly from the data source to the client rather than via the gateway.

This is a highly parallel execution environment designed to allow for optimization of workflows in order to deliver results relying on multiple data sources quickly and conveniently.

7.4 CONCLUSION

There are many aspects to large-scale data-intensive computing and this chapter has only scratched the surface of this vast domain. The digitization of the world around us continues to gather pace. We are reaching a tipping point where traditional methods of data management and analysis will no longer be tenable. This has profound implications for how we manage, understand, and benefit from the data around us. In tandem with the vast increase in data, the world faces an uncertain future with regard to energy production and usage. It seems natural that we should look into low-power options for data-intensive computing, and this chapter has discussed some of the original work in this area currently being undertaken at the University of Edinburgh and elsewhere.

The computational technologies required to optimally store data are only one part of the story, however. We urgently need new data-intensive analysis technologies to make sense of all of the data. The latter part of this chapter therefore focused on the ADMIRE project, which has designed a framework to facilitate next-generation distributed data mining through a carefully designed architecture that separates the data domain experts from the data management experts (the so-called DMI experts and DIDC engineers) such that complexity can be managed and experts encouraged to focus on their specialisms.

The importance of distributed data-intensive computing will continue to expand over the next decade, delivering key insights to scientific research and business analysis. Through the work of projects like ADMIRE, Europe is well-placed to capitalize on this exciting new discipline.

ACKNOWLEDGMENTS

The ADMIRE Project is funded under the European Commission's Seventh Framework Programme through Grant No. ICT-215024 and is a collaboration between the University of Edinburgh, the University of Vienna, Universidad Politécnica de Madrid, Ústav informatiky, Slovenská akadémia vied, Fujitsu Laboratories of Europe, and ComArch S.A.

While preparing the summary of ADMIRE presented here, I have drawn on the work of many project members and many of the project's deliverables. I am grateful to my many colleagues in ADMIRE for all of their hard work on this interesting and engaging research project.

REFERENCES

G. Amdahl. Computer architecture and Amdahl's law. *IEEE Solid State Circuits Society News*, 12(3):4–9, 2007.

A. Hume, L. Han, J. I. van Hemert, et al. ADMIRE—Architecture. Public report D2.1, The ADMIRE Project, Feb 2009.

A. Krause, I. Janciak, M. Laclavik, et al. ADMIRE—Tools development progress report. Deliverable report D5.5, The ADMIRE Project, Aug 2010.

A. Szalay, G. Bell, H. Huang, et al. Low-power Amdahl-balanced blades for data intensive computing. *ACM SIGOPS Operating Systems Review*, 44(1):71–75, 2010.

Chapter 8

A Topology-Aware Evolutionary Algorithm for Reverse-Engineering Gene Regulatory Networks

Martin Swain

Institute of Biological, Environmental and Rural Sciences,
Aberystwyth University, Ceredigion, United Kingdom

Camille Coti

LIPN, CNRS-UMR7030, Université Paris 13, Villetaneuse, France

Johannes Mandel

Roche Diagnostics GmbH, Penzberg, Germany

Werner Dubitzky

School of Biomedical Sciences, University of Ulster, Coleraine, United Kingdom

8.1 INTRODUCTION

This chapter is concerned with modeling and simulating the dynamics of gene regulatory networks (GRNs). Gene regulation refers to processes that cells

Large-Scale Computing, First Edition. Edited by Werner Dubitzky, Krzysztof Kurowski, Bernhard Schott.

use to create functional gene products, in particular proteins, from the information stored in genes. Gene regulation is essential for life as it increases the versatility and adaptability of an organism by allowing it to express or synthesize protein when needed. While aspects of gene regulation are well understood, many open research questions still remain (Davidson and Levin, 2005).

The modeling of dynamic biomedical phenomena such as gene regulation is inspired by the approach taken in physics, where models are frequently constructed to explain existing data, then predictions are made, which again are compared to new data. If a sufficient correspondence exists, it is claimed that the phenomenon has been understood. The ability to construct, analyze, and interpret qualitative and quantitative aspects of gene-regulation models is becoming increasingly important (Hasty et al., 2001). In contrast to static gene expression data, the use of dynamic gene expression data with modeling and simulation helps to determine stable states of gene regulation in response to a condition or stimulus as well as the identification of pathways and networks that are activated in the process.

In this chapter, we study the process of reverse-engineering GRNs from time-series gene expression data sets. The idea is to discover an optimal set of parameters for a computational model of the network that is able to adequately simulate the behavior described by the gene expression data sets. We investigate three different mathematical methods used in computational models that are based on ordinary differential equations. Each system of differential equations has its own characteristics, which lead to biases and artifacts in the network models (Swain et al., 2010). We apply these three mathematical methods to a recent biological data set generated by Cantone et al. (2009) specifically for *in vivo* assessment of reverse-engineering and gene network modeling approaches. The model network presented by Cantone et al. (2009) is a relatively small network. However, such small GRNs can display complex nonlinear behavior due to the various positive and negative feedback loops that may exist between genes in the network.

The mathematical models we investigate require a significant number of parameters to be fine-tuned in order for the models to accurately simulate real biological network behavior. This is a combinatorial optimization problem that often requires considerable computational resources depending on the size and complexity of the network being investigated. We use a reverse-engineering method based on evolutionary algorithms to search through the parameter space and to optimize the values of the parameters in the mathematical models. In the models, there are at least four parameters to be optimized per gene in the network. Each gene may influence the behavior of other genes in the network in complex ways that depend on the nature of the feedback loops. In addition, depending on the mathematical model used, very small variations in a single parameter value can have quite a dramatic effect on the overall network behavior. There may therefore be a great variation in the quantity of computing power required to optimize parameter values in different networks and in different mathematical models. In order to take advantage of available computational resources, we have implemented our parallel evolutionary algorithms using QosCosGrid-OpenMPI (QCG-OMPI).

QCG-OMPI was designed to enable parallel applications to run across widely distributed computational resources owned and secured by different administrative organizations (Coti et al., 2008). The QCG-OMPI middleware is described further in Chapter 9 of this book. An important feature of QCG-OMPI that we have used in this study is its ability to support *topology-aware applications*: Just as complex dynamical systems may consist of many interdependent physical parts, it is also possible that computational models of complex dynamical systems can be decomposed into interdependent functional components that together comprise a topology-aware application; the connections between these components define a specific topology that can be matched to the distributed resources available on a computational grid infrastructure. *Topology-aware middleware*, such as QCG-OMPI, are designed to efficiently support the process of matching topology-aware applications to available networked resources.

This chapter is organized as follows. First, we describe our reverse-engineering methodology in detail, including the different mathematical methods, our optimization approach, and QCG-OMPI. Our results section is divided into two subsections, the first describing how the application scales over distributed resources geographically separated by hundreds of kilometers, and the second demonstrates the accuracy of models reverse engineered using the different mathematical methods. Finally, we conclude with a summary of our main results.

8.2 METHODOLOGY

In this section, we first describe a small synthetic GRN in yeast (Cantone et al., 2009) that we used in our reverse-engineering studies. Subsequently, we outline three systems of equations generally used for modeling GRNs. We then describe QCG-OMPI, which is the novel topology-aware Message Passing Interface (MPI) that we used to implement the communications in our application, and finally, we give the details of our *particle swarm optimization (PSO)* method that we applied to the problem of reverse-engineering GRNs.

8.2.1 Modeling GRNs

The strengths and weaknesses of the mathematical methods are highlighted by using the methods to reverse engineer a real biological network.

8.2.1.1 Case Study: A Yeast Synthetic Network Cantone et al. (2009) synthetically constructed a GRN in yeast to facilitate an *in vivo* assessment of various reverse-engineering and gene network modeling approaches, including approaches based on ordinary differential equations.

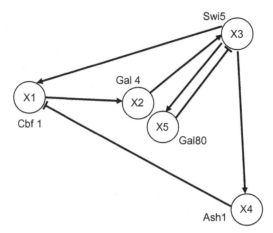

Figure 8.1   *The yeast synthetic network presented by Cantone et al. (2009), with the activating ↑ and repressing ⊥ regulatory interactions.*

Cantone et al. (2009) present expression profiles of the network genes after a shift from a glucose-containing medium to a galactose-raffinose-containing medium; this is called the *switch-on* time series; and after a shift from galactose-raffinose to glucose-containing medium: the *switch-off* time series. In this study, we used the first 100 minutes of these two data sets, excluding the first 10-minute interval during which the washing steps and the subsequent medium shift are performed. After 100 minutes, the biological system is perturbed, and Cantone et al. (2009) used time-delay terms to model this perturbation, which are not present in the basic forms of the equations we study here.

The network includes a variety of regulatory interactions, thus capturing the behavior of larger eukaryotic gene networks on a smaller scale. It was designed to be minimally affected by endogenous genes and to transcribe its genes in response to galactose. While the yeast GRN appears relatively small (see Fig. 8.1), it is actually quite articulated in its interconnections, which include regulator chains, single-input motifs, and multiple feedback loops.

When reverse engineering this GRN, we assume the basic topology of the network; that is, we know which genes are connected, although we do not know if the type of interaction is activating or repressing. The optimization method we use (see Section 8.2.3) then estimates the parameters of a mathematical method. By numerically solving the differential equations with a set of parameters, we can simulate the dynamic behavior of the model, and if the output of our simulations precisely reproduce the time-series data sets recorded by laboratory scientists, then we can claim that our model has reverse engineered the GRN.

The nonlinear differential equations of the three modeling methods investigated in this study describe the mutual activating and repressing influences of genes in a GRN at a high level of abstraction. In particular, it is assumed that the rate of gene expression depends exclusively on the concentration of gene products arising from the nodes (genes) of the GRN. This means that the influence of other molecules (e.g., transcription factors) and cellular processes (translation) is not taken into account directly. Even with these limitations, dynamic GRN models of this kind can be useful in deciphering basic aspects of gene regulatory interactions.

The three methods we have studied have been widely used to model dynamic GRNs. One major advantage of all three methods lies in their simple homogeneous structures, as this allows the settings of parameter discovering software to be easily customized for these structures. In addition, all three modeling methods either already have the potential to describe additional levels of detail or their structures can be easily extended for this purpose.

The three methods describe dynamic GRN models by means of a system (or set) of *ordinary differential equations*. For a GRN comprising N genes, N differential equations are used to describe the dynamics of N gene product concentrations, X_i with $i = 1, \ldots, N$. In all three methods, the expression rate dX_i/dt of a gene product concentration may depend on the expression level of one or more gene products of the genes X_j, with $j = 1, \ldots, N$. Thus, the gene product concentration X_i may be governed by a self-regulatory mechanism (when $i = j$) or it may be regulated by products of other genes in the GRN.

In the following sections, we introduce the three mathematical modeling methods in some detail.

8.2.1.2 The Artificial Neural Network (ANN) Method
Vohradsky (2001) introduced ANN as a modeling method capable of describing the dynamic behavior of GRNs. The way this method represents and calculates expression rates depends on the weighted sum of multiple regulatory inputs. This additive input processing is capable of representing logical disjunctions. The expression rate is restricted to a certain interval where a sigmoidal transformation maps the regulatory input to the expression interval. ANNs provide an additional external input, which has an influence on this transformation in that it can regulate the sensitivity to the summed regulatory input. Finally, the ANN method defines the degradation of a gene product on the basis of standard mass action kinetics.

Formally, the ANN method is defined as

$$\frac{dX_i}{dt} = v_i \cdot f\left(\sum_{j=1}^{N} w_{ij} X_j - \vartheta_i\right) - k_i X_i \quad \vartheta_i, v_i, k_i > 0. \tag{8.1}$$

The parameters of the ANN method have the following biological interpretations:

N: number of genes in the GRN to be modeled; the genes of the GRN are indexed by i and j, where $i, j = 1, \ldots, N$

v_i: maximal expression rate of gene i

w_{ij}: strength of control of gene j on gene i; positive values of w_{ij} indicate activating effects, while negative values define repressing effects

ϑ_i: influence of external input on gene i; influences the responsiveness or speed of the reaction (Vohradsky, 2001) and represents the weighted sum of all inputs, which gives an expression rate equal to half of the maximal transcription rate v_i

f: represents a nonlinear sigmoid transfer function modifying the influence of gene expression products X_j and external input ϑ_i to keep the activation from growing without bounds

k_i: degradation of the ith gene expression product

In the ANN method, the parameters w_{ij} can be intuitively used to define activation and inhibition by assigning positive or negative values, respectively. Here, small absolute values indicate a minor impact on the transcription process (or a missing regulatory interaction), and large absolute values indicate a correspondingly major impact on the transcription process. However, one should be aware that the multiple regulatory inputs required in the case of coregulation can compensate each other due to their additive input processing.

8.2.1.3 TheS-System (SS) Method Savageau (1976) proposed the *synergistic system* or *SS* as a molecular network model. Modeling GRNs with the SS, the expression rates are described by the difference of two products of power law functions, where the first represents the activation term and the second the degradation term of a gene product X_i. This *multiplicative* input processing can be used to define logical conjunctions for both the regulation of gene expression processes and for the regulation of degradation processes. The SS method has no restrictions in the gene expression rates and thus does not implicitly describe saturation.

Formally, the SS method is defined as

$$\frac{dX_i}{dt} = \alpha_i \prod_{j=1}^{N} X_j^{g_{ij}} - \beta_i \prod_{j=1}^{N} X_j^{h_{ij}} \quad \alpha_i, \beta_i > 0, \quad g_{ij}, h_{ij} \in \mathbb{R}. \tag{8.2}$$

The parameters of the SS method have the following biological interpretations:

N: number of genes in the GRN to be modeled; the genes of the GRN are indexed by i and j, where $i, j = 1, \ldots, N$

α_i: rate constant of activation term; in SS GRN models, all activation (upregulation) processes of a gene i are aggregated into a single activation term

β_i: rate constant of degradation term; in SS GRN models, all degradation processes of a gene i are aggregated into a single degradation term

g_{ij}, h_{ij}: exponential parameters called *kinetic order* describing the interactive effect of gene j on gene i; positive values of g_{ij} indicate an activating effect on the expression of gene i, negative values an inhibiting effect; similarly, a positive h_{ij} indicates increasing degradation of the gene product X_i; a negative h_{ij} indicates decreasing degradation

The behavior of SS models is mainly determined by the exponent values g_{ij} and h_{ij}. Similar to the ANN method, high absolute values define strong influence, whereas small absolute values indicate weak influence. However, the dynamics can become particularly complicated when describing inhibiting dependencies using negative exponent values. Here, the effect depends strongly on the actual concentration ranges of the inhibitor, which can introduce singular behavior at near-zero concentrations.

The parameters used in SS models have a clear physical meaning and can be measured experimentally (Hlavacek and Savageau, 1996), yet they describe phenomenological influences, as opposed to stoichiometric rate constants in *general mass action* (*GMA*) systems (Crampin et al., 2004). The SS method generalizes mass action kinetics by aggregating all individual processes into a single activation and a single degradation term (per gene). In contrast, the GMA system defines all individual processes k with $k = 1, \ldots, R$ with the sum of power law functions (Almeida and Voit, 2003).

8.2.1.4 The General Rate Law of Transcription (GRLOT) Method

The GRLOT method has been used to generate benchmark time-series data sets to facilitate the evaluation of different reverse-engineering approaches (Mendes et al., 2003; Wildenhain and Crampin, 2006). GRLOT models multiply individual regulatory inputs. Activation and inhibition are represented by different functional expressions that are similar to Hill kinetics, which allow the inclusion of cooperative binding events (Mendes et al., 2003). Identical to the ANN, the degradation of gene products is defined via mass action kinetics.

Formally, the GRLOT method is defined as

$$\frac{dX_i}{dt} = v_i \prod_j \left(\frac{Ki_j^{n_j}}{I_j^{n_j} + Ki_j^{n_j}} \right) \times \prod_k \left(\frac{A_k^{n_k}}{A_k^{n_k} + Ka_k^{n_k}} \right) - k_i X_i \quad v_i, Ki_j, Ka_j, k_i > 0. \quad (8.3)$$

The parameters of the GRLOT method have the following biological interpretations:

v_i: maximal expression rate of gene i

I_j: inhibitor (repressor) j

A_k: activator k; the number of inhibitors I and the number of activators A can be related to the total number of genes by $I + A \leq N$

Ki_j: concentration at which the effect of inhibitor j is half of its saturation value

Ka_k: concentration at which the effect of activator k is half of its saturation value

n_j, n_k: regulate the sigmoidicity of the interaction behavior in the same way as Hill coefficients in enzyme kinetics

k_i: degradation of the ith gene expression product

The GRLOT method uses two inverse functional forms to describe activating and inhibiting influences, both of which involve two parameters. For example, in the case of activation, each individual dependency is described with a Hill equation where the Hill exponent n_{ij} determines the sigmoid response curve to varying concentration levels of the regulating molecule. Furthermore, the second parameter, Ka_k, allows the concentration of the regulator molecule to be specified, which gives half of its maximal effect on the expression rate. Consequently, high exponent values, together with low Ka_k values, defined for a regulator molecule, indicate that already low concentration levels have a strong activating effect on the transcription process, while, on the other hand, low exponent values in combination with high Ka_k values require high concentration levels of the regulator molecule for effective activation.

Having discussed the mathematical approaches for modeling GRNs, we now outline the computational techniques we developed for reverse engineering these models.

8.2.2 QCG-OMPI

QCG-OMPI (Coti et al., 2008) is an MPI implementation targeted to computational grids, based on OpenMPI (Gabriel et al., 2004) and relying on a set of grid services that provide advanced connectivity techniques. Although QCG-OMPI can be used as a stand-alone component (and this is how we used it for the application presented in this chapter), it is also a key component of the QosCosGrid (QCG) middleware for which it was designed (Kurowski et al., 2009; Bar et al., 2010).

QCG-OMPI is designed to be topology aware. It supports groups of processors that communicate to each other according to different connective topologies. A novel and important feature of QCG-OMPI is that an application is able to adapt to different application topologies according to the set of resources supplied at run time. Also, *any* number of communication layers can be defined. The user has complete flexibility and control in terms of the number of communicators and different grains of parallelism.

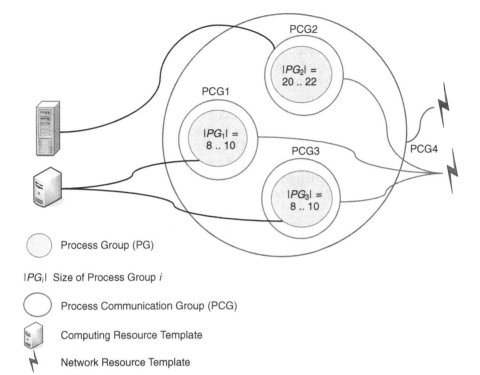

Figure 8.2 *A sample request representation of a topology-aware evolutionary algorithm application.*

The number of groups a process belongs to is called its *depth* and is stored under QCG_TOPOLOGY_DEPTH. The groups defined by the user are called *colors*, and this name is stored under the attribute name QCG_TOPOLOGY_ COLORS. QCG-OMPI provides a QCG_ColorToInt() function to convert alphanumeric colors into integers that can be passed to the MPI_Comm_ split() routine in order to create communicators that fit the topology.

Figure 8.2 depicts an example application based on the evolutionary algorithm software we have used in this chapter. It defines four communication groups or process groups (PG): PG1, PG2, PG3, and PG4. PG4 includes all the processors, while PG1, PG2, and PG3 comprise mutually exclusive subsets. When QCG-OMPI is deployed as a stand-alone component, the details of the process groups may be passed to the application via the environmental variables QCG_TOPOLOGY_DEPTHS and QCG_TOPOLOGY_COLORS. The application can obtain this topology information at any moment during execution, that is, after the MPI library has been initialized and before its finalization.

Additional functionality is available if QCG-OMPI is deployed on the full QCG multiuser infrastructure. In this case, users specify the technical capabilities required for the different process groups, including the properties of the network connecting the different groups. In the QCG, each process group is mapped to a computing resource template that describes the properties of the machines it requires for correct execution. A computing resource template is an instantiation of an XML schema describing the properties of computing resources. To enable a flexible approach to scheduling, and to take advantage of the ability of topology-aware QCG-OMPI applications to adapt to available resources, the computing resource template is described in terms of ranges. For example, the computing resource template requested by both PG1 and PG3 might be specified as follows:

[Clock rate in the range of 2 ... 3 GHz]
[Memory in the range of 1 ... 2 GB]
[Free disk space in the range of 2 ... ∞ GB]

These requirements are then passed to the QosCosGrid Scheduler through the environmental variable `QCG_TOPOLOGY_DEPTHS`.

When used on the QCG infrastructure, *process groups* are arranged in a *process communication group* (*PCG*) topology, where it is assumed that all processes within a PCG would like to have all-to-all interconnections with certain properties. The quantitative properties of these interconnections are specified by means of the network resource templates (a network resource template is an instantiation of an XML schema describing possible network properties).

For instance, PCG1 (one of the small inner circles) contains only processes of PG1, which means that all the processes of PG1 must have all-to-all interconnections as described in the PCG1's network resource template. On the other hand, PCG4 (the large outer circle) contains three process groups—PG1, PG2, and PG3, which means that all the processes of these three process groups must have all-to-all interconnections as described in the PCG4's network resource template. Given any two processes, we can determine the quality of their interconnection by looking into the smallest circle (PCG) that contains these two processes. The network templates are used by the

TABLE 8.1   Array of Colors Corresponding to the Topology Depicted in Figure 8.2

Process Group	PG1	PG1	PG2	PG2	PG3	PG3
Rank	0	1	2	3	4	5
Local rank	0	1	0	1	0	1
Depth	2	2	2	2	2	2
Colors (depth 1)	PCG4	PCG4	PCG4	PCG4	PCG4	PCG4
Colors (depth 2)	PCG1	PCG1	PCG2	PCG2	PCG3	PCG3

QosCosGrid Scheduler, along with the computing resource templates, when allocating network and computing resources for an application.

An example of the topology depicted in Figure 8.2 is given in Table 8.1. This is valid whether QCG-OMPI is used as a stand-alone component or on the QCG infrastructure. In this example, each process group contains two processes. The application's processes, at run time, can obtain the topology description (represented by Table 8.1) by obtaining the appropriate MPI attributes from the run-time environment. For example, the MPI code to fetch these attributes from the run-time environment is

```
/* depths is an array of integers of size NPROCS */
MPI_Attr_get(MPI_COMM_WORLD, QCG_TOPOLOGY_DEPTHS, depths,
&flag);
```

Instructions within the program inform processes what they have to do, regarding their color at a given depth, and what computation pattern they should adopt. Processes can also use colors to adapt their communication patterns. In particular, they can build communicators that fit the requested communication groups using the MPI routine `MPI_Comm_split()`.

The color used by MPI_Comm_split () is the integer translation of the QCG alphanumeric color, given by the `QCG_ColorToInt()` routine. Processes that do not belong to any communication group at a given depth are assigned the `MPI_UNDEFINED` color and create an invalid communicator, as specified by the MPI standard.

QCG-OMPI presents the topology to the application as an array containing (for each process indexed by its rank in the global communicator `MPI_COMM_WORLD`) strings that represent the names of communication groups. Thus, each process in the system is able to determine to which groups it belongs in a very simple way. QCG-OMPI also provides a conversion function to create a unique identifier per communication group name, suitable to be used for the `MPI_Comm_split()` function of the MPI standard, enabling a simple creation of individual communicators per communication group. Using these topology-aware process groups, populations can evolve within each process group and communicate using local communicators, and use the global `MPI_COMM_WORLD` communicator or other user-defined communicators, for cross-group communications only.

The communication middleware of QCG-OMPI takes care of establishing the communication links on demand. This happens transparently for the application developer, even if the peers of the communication are located in different clusters, separated by firewalls, or if the communication involves many peers in many different locations (e.g., when doing a collective operation on a communication group instantiated as a communicator spanning multiple sites). Thus, programming parallel applications in QCG-OMPI remains natural for MPI users, and it becomes simple to discover which resources have been allocated to which communication group (an addition to the MPI standard proposed by QCG-OMPI).

8.2.3   A Topology-Aware Evolutionary Algorithm

Evolutionary algorithms are an approach inspired by biological evolution to iteratively evolve and optimize, according to some criteria, a population of computing objects, where each individual in the population is a potential solution to the problem at hand (Holland, 1992; Bäck, 1996). The individuals of an evolutionary algorithm may be organized around specific population structures or topologies. We distinguish two approaches: island evolutionary algorithms and cellular evolutionary algorithms.

1. *Island evolutionary algorithms* are based on a spatial or topological organization in which a genetic population is divided into subpopulations (islands and regions) that are optimized semi-independently from each other. In this approach, individuals periodically migrate between island subpopulations in order to overcome the problems associated with a single population becoming stuck in a local minima and thus failing to find the global minimum.
2. *Cellular evolutionary algorithm* models are based on a spatially distributed population in which genetic interactions may take place only in the closest neighborhood of each individual (Gorges-Schleuter, 1989). Here, individuals are usually disposed on a lattice topology structure.

PSO is a population-based stochastic optimization technique that shares many similarities with other evolutionary computation techniques (Kennedy and Eberhart, 1995). An appealing feature of PSO is that it requires relatively few parameters when compared to evolutionary algorithms, such as genetic algorithms, that use transformations such as mutation, crossover, and so on, to generate solutions. It is this simplicity that encouraged us to apply it to reverse-engineering GRNs.

In PSO, the population of potential solutions consists of particles that "fly" through the problem space by following a trajectory that is influenced by the current optimal solution, the particle's own best solution, and, in our implementation, the optimal solution of its immediate neighbors.

As with other evolutionary algorithms, PSO is initialized with a group of random solutions and then searches for optima by updating generations. The generations are updated at every iteration by using Equation 8.4 to calculate the particles' current velocities:

$$v_i = v_i + c_1 R_1 \left(r_i^{best} - r_i \right) + c_2 R_2 \left(l^{best} - r_i \right) + c_3 R_3 \left(g^{best} - r_i \right) \qquad (8.4)$$

$$r_i = r_i + v_i. \qquad (8.5)$$

Here R_1, R_2, and R_3 are random numbers between 0 and 1, and the constants c_1, c_2, and c_3 represent learning factors that in PSO algorithms are usually set to 2. The optimal or best current solution or fitness that particle i has achieved is denoted by r_i^{best} and the optimal or best current solution of its local neighbors is denoted by l^{best}. Finally, the optimal or best current value obtained by any particle in the subpopulation is given by g^{best}. Periodically, using the message passing functions implemented by QCG-OMPI, the particles communicate their current optimal or best solutions, r_i^{best}, with the rest of the population and receive values for l^{best} and g^{best}.

The PSO algorithm is one of a number of evolutionary algorithms that we implemented in a program called CellXPP. CellXPP is written in C, and while the QCG-OMPI routines are handled by the C program, the fitness function uses a Python interface to call XPPAUT,[1] a program maintained by G. Bard Ermentrout to solve the differential equations. We used XPPAUT because it is a well-established piece of software with a large user community and, more importantly, because it provides a command-line interface and is easy to install on a variety of different machine architectures. The Python interface was used to convert the current parameter set of a particle into an execution file (with the correct set of differential equations for XPPAUT) and then process the output of XPPAUT using Equation 8.6 to calculate a fitness value. Python was used because of its scripting capabilities that allowed us to quickly experiment with new sets of differential equations and different approaches to calculate the fitness.

The application consists of two layers or PCGs that combine features of both cellular and (cooperative) island evolutionary algorithms, thus giving the application a two-layer topology. In this setup, the algorithm is aware of the two layers; hence, we call it a topology-aware evolutionary algorithm. More frequent communications, at every iteration, take place in the lower layer (cellular aspect of the combined evolutionary algorithm approach), and less frequent communications, perhaps every 100 iterations, take place in the higher layer (insular aspect). These are shown in Figure 8.3 with a ring topology used for communications between island subpopulations, while in the cellular layer, individuals communicate with their immediate neighbors.

In Figure 8.3, the evolutionary algorithm is composed of three subpopulations, which may be distributed over any number of physical locations. However, in general, the algorithm can divide the population into one or more subpopulations. A subpopulation may run over multiple physical locations, and it is also possible for multiple subpopulations to run at a single physical location. The exact distribution of the population is left to the user to decide and may be specified at run time. The topology-aware functionality of QCG-OMPI is then able to adapt the application to the specified communication topology. Normally, due to the relatively frequent communication patterns taking place within subpopulations, subpopulations are most efficiently executed on one computational cluster at a single physical location.

Time-series gene expression data form the input to our reverse-engineering process. We aim to identify model parameters for which the discrepancy

---

[1] http://www.math.pitt.edu/~bard/xpp/xpp.html.

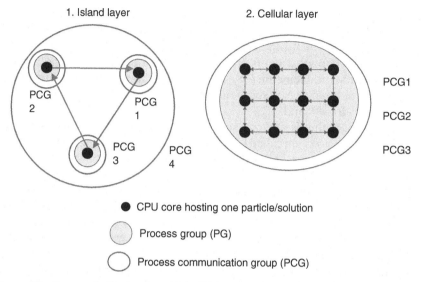

Figure 8.3 *Communication topology of the PSO algorithm. The arrows show the communication patterns in each process communication group: Communications in the island layer can be mapped onto a ring topology, while communications in the cellular layer can be mapped onto a spherical topology where each process communicates to its neighbors.*

between the experimental data and the data predicted by the model is minimal (Wessels et al., 2001). This leads to an optimization problem in which the following function is to be minimized:

$$f = \sqrt{\sum_{t=1}^{T}\sum_{k=1}^{N}\left(X_k(t) - \hat{X}_k(t)\right)^2}, \qquad (8.6)$$

where $X_k(t)$ denotes the experimentally *observed* gene product concentration level of gene k at time point t; $\hat{X}_k(t)$ denotes the gene product concentration level of gene k at time point t *predicted* by the dynamic GRN model; N denotes the number of genes in the network; and T represents the number of sampling points of observed data.

The parameter space searched for the three mathematical methods described in Section 8.2.1 was defined as follows:

- **ANN Method:** w_{ij} from −15.0 to 15.0, v_i from 0.0 to 3.0, ϑ_i from 0.0 to 7.0, and k_i from 0.0 to 2.0
- **SS Method:** g_{ij} and h_{ij} from −3.0 to 3.0, and α_i and β_i from 0.0 to 15.0
- **GRLOT Method:** I_j, from −5.0 to 5.0, A_k from 0.0 to 7.0, Ki_j and Ka_k from 0.0 to 40.0, and k_i from 0.0 to 2.0

The topology awareness of the evolutionary algorithm could be extended to include a third topology layer that would be used if the fitness calculation had to be parallelized. For example, if the machines allocated have multiple cores or multiple processors, then one way of implementing the third topology layer involves placing all available processes from a machine into a group. The topology-aware algorithm would then be able to adapt to each machine, depending on the number of cores and processors that it possesses. The three layers of the topology-aware algorithm would then include parallelization at the machine level, parallelization at the cluster level, and parallelization at the cross-cluster or grid level.

8.3 RESULTS AND DISCUSSION

First, we discuss some performance characteristics of the GRN reverse-engineering application in terms of its computational efficiency and ability to scale over distributed resources, then we show the results of our reverse-engineering experiments with the Cantone et al. (2009) yeast regulatory network and the three mathematical methods.

8.3.1 Scaling and Speedup of the Topology-Aware Evolutionary Algorithm

Using the topology-aware evolutionary algorithm described in Section 8.2.3, we have conducted a number of computational studies on Grid'5000, which is an academic, nationwide experimental grid dedicated to research (Cappello et al., 2005). In these studies, we used QCG-OMPI as a stand-alone component, without any other features of the QCG middleware. For the experiments we performed, we were allocated resources for which we were the only user.

We used two sites, located in the cities of Orsay and Bordeaux in France. These locations are physically separated by several hundred kilometers. The specifications of the clusters are as follows:

- In Orsay, a cluster called GDX was used. It is based on AMD Opteron machines with two single-core CPUs running at 2.4 GHz and with 2.0-GB memory. For the Orsay cluster experiments, we had 120 cores available in 60 machines.
- In Bordeaux, a cluster called Bordereau was used. It is based on AMD Opteron machines with dual-processor dual-core CPUs running at 2.6 GHz and 4.0-GB memory. For the Bordeaux cluster experiments, we had 120 cores available in 30 machines.

The nodes within each cluster were interconnected by a Gigabit Ethernet switch, and the clusters were connected by the Renater French Education and Research Network using a 10-Gb/s dark fiber.

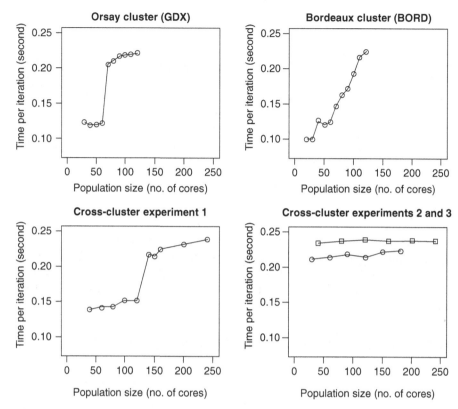

Figure 8.4 *Speedup results of the evolutionary algorithm using different clusters and computational resources. Note that the number of cores corresponds to the population size and hence the number of fitness calculations performed in an iteration. The graphs are discussed in the main text.*

The evolutionary algorithm performs one fitness calculation per CPU core. If the algorithm is scaling well, as the number of cores increases, we would hope to see the time required to perform a fitness calculation to remain constant—so, 200 cores perform 200 fitness calculations in the same time as 40 cores perform 40 fitness calculations.

Figure 8.4 depicts the results of the five experiments using the two Grid'5000 clusters. It shows the speedup of the evolutionary algorithm execution and depicts the average time to execute an iteration of the evolutionary algorithm when running on different clusters with different topologies. At each iteration, an individual (or a process running on a different CPU core) performs a fitness calculation and communicates the results to its neighboring processes. As the number of cores increases, so does the size of the population, which results in an increase in the number of fitness calculations performed at each iteration.

In Figure 8.4, the graphs labeled "GDX" and "BORD" depict execution times of runs performed on a single cluster, corresponding to the Orsay cluster

(GDX) and the Bordeaux cluster (BORD), respectively. The other two graphs (labeled "Cross-cluster experiment 1," "Cross-cluster experiment 2," and "Cross-cluster experiment 3," respectively) show three cross-cluster runs performed on both clusters (GDX and BORD).

For the GDX simulations, we initially ran just 1 process per machine, up to 60 processes; then, we used both CPUs on each machine, up to the full number of 120 CPUs. There is a noticeable increase in execution time at the transition from one to two CPUs per machine. Up to 60 CPUs, the speedup is linear, and above 60 CPUs, the speedup is also approximately linear but with a greater time required to perform a full iteration.

For the BORD simulations, we initially restricted processes to run on just 1 core per machine (30 cores in total). Then, we increased this to 2 (from 30 to 60 cores), then to 3 (from 60 to 90 cores), and finally, we used all 4 cores available in each machine. Here, we observed a sudden increase in execution time at the transition from one to two cores per machine—here, the two cores are on separate CPUs within the same machine. However, when progressing from two to three or four cores, the execution time increases according to the number of cores used—in this case, the speedup is not linear. When we increased the number of processes per machine, we kept the number of network interface cards constant. We speculate that there is congestion at the level of network interface cards because several processes are trying to communicate through the network using a single network interface card. Thus, the communication time within a machine is greater when running several processes per machine.

For Cross-cluster experiment 1, we performed a cross-cluster run by trying to minimize the number of processes run on each machine. This experiment follows a similar pattern to the runs on GDX and BORD: Initially, one process is run on a machine using one CPU on each machine on GDX and one CPU on BORD (from 30 to 90 cores). Next, we used two BORD CPUs (from 90 to 120 cores) followed by both the CPUs for the GDX machines (from 140 to 160 cores). And for the final two points where population size (number of cores) is changed from 210 to 240 cores, we used all the available cores in the machines. The shape of the graph is similar to that of the individual runs, particularly the GDX run with its sudden increase in execution time when both CPUs are used in each machine. Overall, the execution times are slightly longer when running across both clusters; this may be due to the increased communication overhead between the different physical locations.

For the Cross-cluster experiments 2 and 3, we used the maximum available cores and/or CPUs in each machine; that is, for Cross-cluster experiment 2, we used all the available cores per machine on GDX and two cores per machine on BORD (this is the lower line in the figure); and for Cross-cluster experiment 3, we ran two processes per machine on GDX and four processes per machine on BORD (this is the top curve in the bottom-right graph in Fig. 8.4). These simulations show that the speedup is approximately linear. Since the speedup is linear, this means that the communication pattern is scalable; that is, there is no bottleneck. This implies that our cellular communication pattern is a good architecture for this infrastructure, and that it does not suffer from congestion, which might be a problem for more common parallelization schemes based on master–worker communication patterns.

It is clear for this application that the cross-cluster communications (hundreds of kilometer distance between Orsay and Bordeaux) are not a limiting factor. Instead, the main limitation arises from the use of Python scripts to interface with the XPPAUT software through system calls. This is due to the relatively inefficient operations caused by running multiple processes on the individual machines—when several processes are running on the same machine, a situation of concurrency between processes on disk accesses can be created. This hinders the sequential parts of the parallel algorithm and thus increases the total execution time.

8.3.2 Reverse-Engineering Results

The 10 graphs in Figures 8.5 and 8.6 show the gene expression data of the five genes in the Cantone network, *CBF1, GAL4, SWI5, GAL80,* and *ASH1,* along

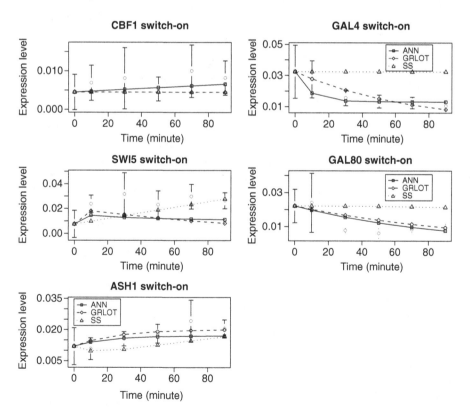

Figure 8.5 *Reverse-engineering results for switch-on data from Cantone et al. (2009). In the graphs, the points with error bars depict experimental data and the lines represent the models.*

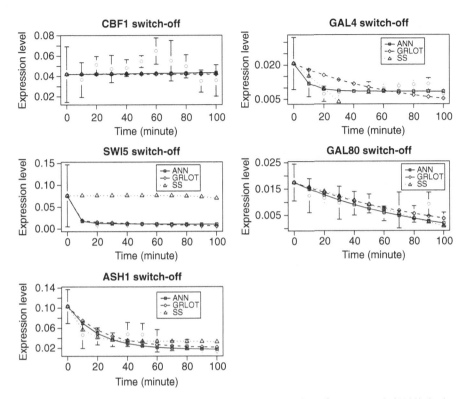

Figure 8.6 *Reverse-engineering results for switch-off data from Cantone et al. (2009). In the graphs, the points with error bars depict experimental data, and the lines represent the models.*

with the dynamics predicted by the three methods. The error bars shown are for the experimentally derived data of Cantone et al. (2009). The ANN and GRLOT methods tend to provide a good match to the experimental data, with the ANN method giving the best results in this test. The predictions made by the SS method are clearly worse than the other two methods.

The SS method couples unrestricted expression rates with input processing based on multiplied exponentials, which, in the experiments we have explored in this chapter, frequently leads to extreme sensitivity and unstable dynamics. Indeed, it is common that the parameter values automatically generated by the evolutionary algorithm for the SS method do not give any solution. This is because extremely large exponential terms in the equations lead the integrator to quickly fail. As a result, the evolutionary algorithm tends toward solutions where the influence of the power terms is minimized by setting the values of their parameters to values close to zero, giving linear graphs. However, as genes may be close to a steady state of expression for much of the time series, a linear solution may be a reasonably accurate solution and thus does not significantly impact on the overall model output.

The three methods are designed to be flexible enough to find phenomenological expressions for their approximations of experimentally observed behavior (Vohradsky, 2001; Crampin et al., 2004). In contrast to GRN models based on the SS and GRLOT methods, where regulatory inputs are exponentiated and multiplied, GRN models based on the ANN method weight and sum multiple regulatory inputs. An attraction of the SS and GRLOT methods is that their input processing accords better than that of the ANN method with the fundamentals of reaction kinetics and collision theory (chemical reaction rates correlate with the multiplied concentrations of the reactants). However, none of the three methods follow strictly the chemical reaction processes because a single kinetic equation typically comprises several processes. An advantage of the ANN method is that its approach of weighting and summing regulatory inputs is relatively robust and avoids the extreme sensitivity of the SS method. For example, when the evolutionary algorithm randomly selects parameters for the ANN method, they almost always lead to solvable equations. Of the three methods, it is our experience that the ANN method tends to require relatively less computation to achieve an accurate solution and is therefore to be preferred over the GRLOT and SS methods. A more detailed comparison of the three methods is given in Swain et al. (2010).

8.4   CONCLUSIONS

QCG-OMPI is an effective and efficient framework for distributed evolutionary algorithms. The topology awareness of QCG-OMPI simplifies the software development process for the evolutionary algorithm, and, although the programmer must define the functionality of the communication groups within the code, there is no need to define the number or size of the groups as these are allocated automatically at run time.

Distributed evolutionary algorithms often combine a number of different topologies, such as the master–worker topology or the topologies we have explored in this chapter. QCG-OMPI makes it easy to define different communication groups in a flexible way so that an evolutionary algorithm can automatically adapt to different collections of computing resources. As a result, programmers can exploit distributed resources and parallel topologies to develop novel approaches to reverse-engineering GRNs. For example, different island populations might use different mathematical methods in order to overcome the biases and weaknesses of generating models with just a single method. While complete sets of parameters are not transferable between different systems of equations, there are useful results that may be transferred. For instance, if the topology of the network is unknown, then one task of the reverse-engineering process involves predicting which genes are *not* interacting with each other, which is modeled by putting the weights w_{ij} to zero in the ANN method, or the exponents n_{ij} to zero in the GRLOT method. Results such as these are easily transferable between the mathematical methods.

Even for very small gene networks, considerable amounts of highly time-resolved data, with many sampling points based on multiple perturbations and experimental conditions, may be necessary to produce reliable dynamic GRN models. Our studies suggest that mathematical methods used to model GRNs are not equivalent, and because of considerable differences in the way the methods process inputs, the models they create may vary considerably in terms of accuracy. Furthermore, as the complexity in terms of the number of genes and regulatory interactions increases, the computational complexity and computing power required to reverse engineer dynamic GRN models becomes nontrivial, requiring nonstandard computing solutions such as clusters, super-computers, or other large-scale computing solutions. QCG-OMPI and the software discussed in this chapter are an attractive solution to meet these challenges.

ACKNOWLEDGMENTS

The work reported in this chapter was supported by EC grants DataMiningGrid IST FP6 004475, QCG IST FP6 STREP 033883, and ICT FP7 MAPPER RI-261507. Some of the presented experiments were carried out using the Grid'5000 experimental test bed, being developed under the INRIA ALADDIN development action with support from the Centre National de la Recherche Scientifique (CNRS), Reseau National de Télécommunication pour la Technologie, l'Enseignement et la Recherche (RENATER), and several universities as well as other funding bodies (see https://www.grid5000.fr).

REFERENCES

J. S. Almeida and E. O. Voit. Neural-network-based parameter estimation in S-system models of biological networks. *Genome Informatics*, 14:114–123, 2003.

T. Bäck. *Evolutionary Algorithms in Theory and Practice: Evolution Strategies, Evolutionary Programming, Genetic Algorithms*. Oxford, UK: Oxford University Press, 1996.

P. Bar, C. Coti, D. Groen, et al. Running parallel applications with topology-aware grid middleware. In *Fifth IEEE Int'l Conference on eScience, Oxford, UK, December 9–11, 2009*, pp. 292–299, 2010.

I. Cantone, L. Marucci, F. Iorio, et al. A yeast synthetic network for in vivo assessment of reverse-engineering and modeling approaches. *Cell*, 137(1):172–181, 2009.

F. Cappello, E. Caron, M. Dayde, et al. Grid'5000: A large scale and highly reconfigurable grid experimental testbed. In *Proc. of the 6th IEEE/ACM Int'l Workshop on Grid Computing (SC'05)*, pp. 99–106, Seattle, WA: IEEE/ACM, 2005.

C. Coti, T. Herault, S. Peyronnet, et al. Grid services for MPI. In *8th Int'l Symposium on Cluster Computing and the Grid (CCGRID'08)*, pp. 417–424, Washington, DC: IEEE Computer Society, 2008.

E. J. Crampin, S. Schnell, and P. E. McSharry. Mathematical and computational techniques to deduce complex biochemical reaction mechanisms. *Progress in Biophysics and Molecular Biology*, 86(1):77–112, 2004.

E. Davidson and M. Levin. Gene regulatory networks. *Proceedings of the National Academy of Sciences of the United States of America*, 102(14):4935–4935, 2005.

E. Gabriel, G. E. Fagg, G. Bosilca, et al. Open MPI: Goals, concept, and design of a next generation MPI implementation. In *Proc. of the 11th European PVM/MPI Users' Group Meeting*, pp. 97–104, Budapest, Hungary, September 2004.

M. Gorges-Schleuter. Asparagos an asynchronous parallel genetic optimization strategy. In *Proc. of the 3rd Int'l Conference on Genetic Algorithms*, pp. 422–427, San Francisco, CA: Morgan Kaufmann Publishers Inc., 1989.

J. Hasty, D. McMillen, F. Isaacs, et al. Computational studies of gene regulatory networks: In numero molecular biology. *Nature Reviews Genetics*, 2(4):268–279, 2001.

W. S. Hlavacek and M. A. Savageau. Rules for coupled expression of regulator and effector genes in inducible circuits. *Journal of Molecular Biology*, 255(1):121–139, 1996.

J. H. Holland. *Adaptation in Natural and Artificial Systems: An Introductory Analysis with Applications to Biology, Control and Artificial Intelligence*. Cambridge, MA: MIT Press, 1992.

J. Kennedy and R. Eberhart. Particle swarm optimization. In *Proc. of the IEEE Int'l Conference on Neural Networks. IV*, pp. 1942–1948, 1995.

K. Kurowski, W. Back, W. Dubitzky, et al. Complex system simulations with QosCosGrid. In K. Kurowski et al., editors, *ICCS '09: Proc. of the 9th Int'l Conference on Computational Science*, pp. 387–396, Berlin: Springer-Verlag, 2009.

P. Mendes, W. Sha, and K. Ye. Artificial gene networks for objective comparison of analysis algorithms. *Bioinformatics*, 19(90002):122–129, 2003.

M. A. Savageau. *Biochemical Systems Analysis: A Study of Function and Design in Molecular Biology*. Reading, MA: Addison-Wesley, 1976.

M. T. Swain, J. J. Mandel, and W. Dubitzky. Comparative study of three commonly used continuous deterministic methods for modeling gene regulation networks. *BMC Bioinformatics*, 11(459):1471–2105, 2010.

J. Vohradsky. Neural network model of gene expression. *The FASEB Journal: Official Publication of the Federation of American Societies for Experimental Biology*, 15(3):846–854, 2001.

L. Wessels, E. van Someren, and M. Reinders. A comparison of genetic network models. In *Proc. of the Pacific Symposium on Biocomputing*, pp. 508–519, 2001.

J. Wildenhain and E. J. Crampin. Reconstructing gene regulatory networks: From random to scale-free connectivity. *IEE Proceedings of Systems Biology*, 156(4):247–256, 2006.

Chapter **9**

QosCosGrid e-Science Infrastructure for Large-Scale Complex System Simulations

Krzysztof Kurowski, Bartosz Bosak, Piotr Grabowski,
Mariusz Mamonski, and Tomasz Piontek
Poznan Supercomputing and Networking Center, Poznan, Poland

George Kampis
Collegium Budapest (Institute for Advanced Study), Budapest, Hungary

László Gulyás
Aitia International Inc. and Collegium Budapest
(Institute for Advanced Study), Budapest, Hungary

Camille Coti
LIPN, CNRS-UMR7030, Université Paris 13, Villetaneuse, France

Thomas Herault and Franck Cappello
National Institute for Research in Computer Science and Control
(INRIA), Rennes, France

9.1 INTRODUCTION

Grids and clouds could be viewed as large-scale computing systems with considerable levels of hardware resources but lacking many of the the features that make supercomputers so powerful. In particular, grids and clouds usually

Large-Scale Computing, First Edition. Edited by Werner Dubitzky, Krzysztof Kurowski, Bernhard Schott.

do not provide sophisticated support for parallel and multiphysics applications with significant interprocess communication requirements. Connected via local and wide area networks, such big computing infrastructures typically rely on an opportunistic marshalling of resources into coordinated action to meet the needs of large-scale computing applications. Both grids and clouds are often presented as a panacea for all kinds of computing applications, including those that require supercomputing-like environments. However, this vision of grids or clouds as virtual supercomputers is unattainable without overcoming their performance, coallocation, and reliability issues. The demanding nature of scientific simulations requires a new e-infrastructure that is able to simultaneously manage heterogeneous resources, such as computing resources, storage, and networks to guarantee the level of quality of service demanded by end users and their applications, especially a large number of legacy applications designed to run in parallel. To meet the requirements of large-scale complex simulations, we have built a system capable of bringing supercomputer-like performance to advanced applications, including sophisticated parameter sweep experiments, workflow-intensive applications, and cross-cluster parallel computations. Our system (called QosCosGrid [QCG]) consists of a collection of middleware services and external application tools and was first introduced by Kurowski et al. (2009). QCG is designed as a multilayered architecture that is capable of dealing with computationally intensive large-scale, complex, and parallel simulations that are usually too complex to run within a single computer cluster or machine. The QCG middleware enables computing resources (at the level of processor cores) from different administrative domains (ADs) to be combined into a single powerful computing resource via the Internet. Clearly, bandwidth and latency characteristics of the Internet may have an effect on overall application performance of QCG-enabled applications. However, the ability to connect and efficiently control advanced applications executed in parallel over the Internet is a feature that is highly appreciated by QCG users.

QCG could be viewed as a *quasi-opportunistic supercomputer* whose computational performance exceeds the power offered by a single supercomputing or data center (Kurowski et al., 2010). Nowadays, it is more common for complex system simulations to rely on supercomputers because of the high data volume and computing requirements of the individual computations, but also because of the high communication overhead between the computation tasks on individual elements. However, dedicated supercomputers for such calculations are expensive to acquire and maintain. As a consequence, many organizations do not have access to supercomputing facilities and rely local computing resources. Recently, local computing clusters and other multicore and multimachine systems have become the technology of choice for many complex system modelers. However, with the advent of flexible modeling tools, complex system simulations have become even more difficult to manage. As a result, local clusters are increasingly inadequate to satisfy the required computing and communication needs. QCG aims to address this gap by facilitating supercomputer-like performance and structure through efficient cross-cluster computations. Thus, new middleware services and application tools for end users were developed and integrated to narrow the gap and to realize large-scale parallel and distributed computing e-infrastructures. The QCG implementation comprises a comprehensive framework for metascheduling and managing topology-aware complex system applications. This framework includes pluggable components that carry out the usual scheduling operations, including the assignment of parallel processes on a time axis, clustering of resources, and matching application requests[1] and available resources across geographic dispersed locations. The QCG framework is highly flexible as it is composed of pluggable components that can be easily modified to support different scheduling and access policies to better maximize a diversity of utility functions. Furthermore, the framework exploits novel algorithms for topology-aware coallocations that are required by parallel programming and execution setups in production-level high-performance computing (HPC) environments, such as the Message Passing Interface (MPIs), ProActive, or their hybrid extensions linking programming models like OpenMP or CUDA.

This chapter focuses on two enhanced and widely used parallel computing environments: QosCosGrid-ProActive (QCG-ProActive) and QosCosGrid-OpenMPI (QCG-OMPI). QCG-OMPI[2] is designed to enable parallel applications to run across geographically widely distributed computational resources owned and secured by different administrative organizations (Agullo et al., 2011). QCG-ProActive was successfully integrated with the Repast Suite, a very popular Java-based, agent-based modeling and simulation platform (Gulyás et al., 2008). A new version of the C++ implementation of Repast for supercomputing environments (called Repast HPC[3]) will be supported by QCG.

The remainder of this chapter is organized as follows. Section 9.2 presents a short overview of validation scenarios; it also discusses the main requirements of complex systems and other demanding parallel applications and classifies these into "templates." Section 9.2 gives a technical overview of QCG-ProActive and QCG-OMPI. In Section 9.4, we introduce the QCG middleware services and their key capabilities relevant to end users. A number of useful libraries and frameworks for application developers and resource providers using QCG are described in Section 9.5. Section 9.6 presents additional Web-based monitoring and troubleshooting tools for QCG, which are already available on some production-level computational clusters and supercomputers. Finally, Section 9.7 concludes this chapter and discusses future development plans.

9.2 DISTRIBUTED AND PARALLEL SIMULATIONS

In this section, we briefly introduce the main requirements of a variety of complex system applications. To classify frequently occurring application

---

[1] We have categorized these into six generic templates; see Section 9.2.
[2] More details on the QCG-OMPI middleware are presented in Chapter 8 of this volume.
[3] Chapter 5 presents an overview of the Repast HPC framework and its implementation.

patterns into communication templates, Gulyás et al. (2008) studied the inter-action topologies of a wide range of complex system simulations. The proposed communication templates were used to inform the design and implementation of the QCG middleware. The QCG middleware controls a hierarchical struc-ture of scheduler and map application requests to distributed computational resources and networks to enable efficient processing. Each template may be accompanied by one or more template implementations and parameters that determine how the information in the templates is to be used by the QCG middleware. This feature is specifically designed to meet the needs of complex system modelers. The modelers take advantage of this by identifying the com-munication class their models belong to and then populate the templates with the details of their specific simulation models and tasks. While this approach may not achieve the most efficient distributed implementation, it is likely to reduce the implementation effort that would otherwise be required to realize many simulations.

Two special application patterns may be distinguished. In the first, the dependencies among components do not follow a predetermined structure (i.e., they form a uniform random distribution) and change regularly over time (e.g., dependencies are resampled prior to each update). The second special case is when no communication occurs among complex system components.

One might argue that these are overly simplified examples and that a col-lection of components (as in the second example) may not qualify as a system at all. However, we believe this is merely a question of the level of abstraction adopted. In the first scenario, which is not as uncommon as it may seem, a distributed parameter space search is arguably the most adequate implemen-tation strategy, which is just an example of our second scenario: the individual simulation runs can be viewed as noninteracting components. On the other hand, more sophisticated parameter space search methods introduce depen-dencies among individual runs by determining which parameter combination to explore next on the basis of the results computed previously (i.e., sampling in more "turbulent" parameter regions). In this case, a parameter space search becomes a nontrivial complex system again, worthy of dependency analysis in its own right. Dealing with more complex cases, our first observation is that static communication patterns allow for the direct application of distribution algorithms. Therefore, we handle these cases separately from the dynamic topologies. Next, we point out that the dependencies of the update functions may be dependent on the components' states. In many cases, it is possible to project the components' state information to a metric space and update depen-dencies based on distance in this space. For example, if components are agents moving in a space (in computational models often realized as a two-dimensional lattice), then each agent state will (among other things) include the coordi-nates of the agent. If in the model the agents interact only with the agents in their vicinity, then the update dependencies of the components are distance dependent. This spatial property of a complex system simulation, if present, may be successfully exploited in determining the partitions of a distributed

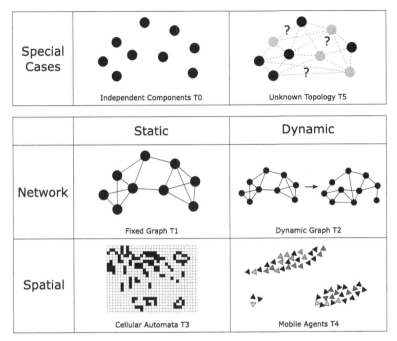

Figure 9.1 *QosCosGrid communication templates that help users to understand and classify main complex system requirements for large-scale simulations.*

implementation. It is worth noting, however, that complex systems may have such a spatial property implicitly. For example, a social system where people are likely to interact with like-minded partners may have this property where the natural metric space is an abstract similarity space. Based on the observations above, we propose five plus one *communication templates* that, as we believe, represent commonly occurring classes of complex system simulations (Fig. 9.1).

Clearly, it is possible to identify many more templates or to refine the presented classification scheme by introducing subclasses. However, we believe the scheme of six main template categories (Fig. 9.1) is sufficient to realize a wide range of complex system simulations and to implement a flexible approach to support efficient distributed computing strategies.

Template 0 (T0) of our classification is the case where no interaction occurs among components. Here, the component partitioning is only constrained by load balancing considerations.

Template 5 (T5), the other extreme, is the case with random or unpredictable interaction among components. In general, we are not able to predict any communication topology in advance so we recommend to run such components independently. The four remaining cases are created at the intersection of the spatial/nonspatial and static/dynamic properties.

Template 1 (T1) (static networks) describes a nonspatial system with a static communication pattern. It is assumed that the exact communication pattern can be extracted from the system, or that it is defined explicitly. To the thus defined communication graph, a variety of graph partitioning algorithms can be applied, in particular, approaches proposed by Fjallstrom (1998).

Template 2 (T2) (dynamic networks) introduces dynamism into Template 1. The assumption about the existence of a communication graph is maintained, but, in contrast to Template 5, it is assumed that the changes and their frequency are defined by a graph transition function that provides enough information to predict in advance communication requirements among distributed components. (Sometimes, it may be sufficient to know that the level of change in the communication graph is low, such that it is sufficient to repartition nodes at regular intervals, at every 10,000 time steps, for instance.) The implementation approach we propose for such systems is the regular application of classic graph repartitioning algorithms, that is, graph partitioning algorithms that attempt to improve on an existing partition (Barnard, 1995).

Template 3 (T3) (static spatial systems) moves away from Template 1 along the other axis. It maintains the assumption about a static communication pattern but requires the spatial property. Prime examples of such systems are cellular automata. Template 3 is the pattern that is geared toward a distributed implementation; such implementations may be based on methods and algorithms developed for distributed cellular automata (Mazzariol et al., 2000).

Finally, *Template 4* (T4) (dynamic spatial systems) assumes a spatial system in which the communication pattern evolves over time. One example of such a system is the case discussed earlier in which agents move in space and communicate with other agents in their immediate vicinity. To realize distribution computing strategies for complex systems belonging to this template, we point to algorithms specially developed for such systems using buffering and messaging solutions and ways of predicting the speed of spatial movement (Scheutz and Schermerhorn, 2006). It is worth pointing out that the fundamental assumption of this template and thus a key to the successful implementation of these solutions is that changes in spatial positions are slow relative to the frequency of state updates.

9.3 PROGRAMMING AND EXECUTION ENVIRONMENTS

The main goal in the development of the QCG middleware was to provide a flexible, efficient, and secure distributed computing system that is able to deal with large-scale simulations over distributed computing resources connected via local and wide area networks (in particular via Internet connections). To inform the development of the QCG middleware and to facilitate its testing and validation, a number of concrete resource-demanding complex system simulations have been identified and classified into communication templates (Section 9.2). The initial set of use cases included the living simulation, evolutionary computation, and agent-based modeling. Recently, we added new requirements arising from various multiscale and multiphysics use cases to be support by QCG, for example, the multiscale modeling in computational biomedicine (Sloot and Hoekstra, 2010).

From a development perspective, the applications were grouped into two classes: (1) Java applications taking advantage of the ProActive library as the parallelization technology and (2) applications based on ANSI C or similar codes, which rely on the message passing paradigm. Based on these groups, QCG was designed to support two parallel programming and execution environments, namely, QCG-OpenMPI (aiming at C/C++ and Fortran parallel applications developers) and QCG-ProActive (aiming at Java parallel application developers).

In this section, we briefly describe those two programming and execution environments together with a number of useful features that make them easy to use and powerful for end users concerned with large-scale parallel simulations of complex systems. All the presented technologies and integration efforts described in the next two subsections were needed to faciliate cross-cluster execution of advanced applications in firewall-protected and Network Address Translation (NAT) environments. Next, we also present our new approach to a common problem in many grids and clouds regarding the simultaneous access and synchronization of a large number of computational resources. We also provide an overview of the innovative cross-cluster deployment protocol developed within QCG, which is designed to simplify the management of complex parallel processes that are organized in groups and hierarchies. Even though the QCG-managed cross-cluster parallel executions are limited to the two enhanced parallel environments, other advanced management capabilities for applications are supported as well, including support for workflows or parameter sweep jobs.

9.3.1 QCG-OMPI

MPI is a de facto a standard in the domain of parallel applications. MPI provides end users with both the programming interface consisting of simple communication primitives and the environment for spawning and monitoring MPI processes. A variety of implementations of the MPI standard are available (both as commercial and open source). QCG uses the OpenMPI implementation of the MPI 2.0 standard. Of key importance are the intercluster communication techniques that deal with firewalls and NAT. In addition, the mechanism for spawning new processes in OpenMPI is integrated in the QCG middleware. The extended version of the OpenMPI framework is referred to as QCG-OMPI (Coti et al., 2008). The QCG-OMPI is as follows:

1. Internally, QCG-OMPI improves the MPI library through a variety of connectivity techniques to enable direct connections between MPI ranks that are located in remote clusters, potentially separated by firewalls.

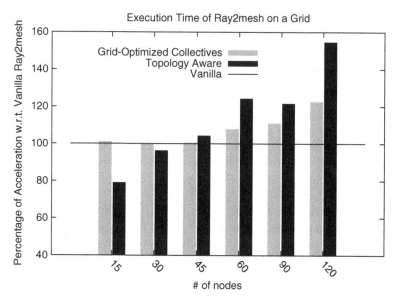

Figure 9.2 *QosCosGrid-OpenMPI (QCG-OMPI) performance on test-bed resources.*

2. Mechanisms going beyond the MPI standard were realized to accommodate QCG's semiopportunistic approach; this was achieved by a new interface to describe the actual topology provided by the Metascheduler.

3. MPI collective operations were upgraded to be hierarchy aware and were optimized for the grid.

We briefly present in this section each of these extensions and some results that demonstrate the performance achievable with QCG-OMPI.

Agullo et al. (2010) present two MPI applications that have been ported to the QCG approach using QCG-OMPI. The first is a simple ray tracing application (Ray2mesh) based on a hierarchical master–worker scheme. Results from performance tests are depicted in Figure 9.2 and are contrasted with the performance of non-QCG implementations of this application. Using grid-aware collective operations is a simple approach for porting an application to the grid. Ray2mesh uses collective operations to communicate parameters and collect partial results from the computing nodes. Figure 9.2 illustrates that the performance obtained through Ray2mesh using grid-optimized collectives (gray bars) is consistently outperforming the Vanilla implementation; the performance gain increases as the number of processes increase (20% with 120 processes).

Another key element of porting an application to the grid requires the adaptation of communication and computation patterns that fit the underlying topology (black bars). The performance improvement at a larger number of nodes or processors is outperforming topology-aware collective communications (55% with 120 processes). We can see that small-scale executions show lower performance than the Vanilla implementation. The reason for this is that for a given number of processes, the grid-enabled implementation dedicates more processes to control and scheduling (i.e., master processes) than the Vanilla implementation. As a consequence, fewer processes are available for computation (i.e., worker processes). However, it would be possible to run worker processes on the same nodes as master processes since the latter are carrying out input/output (I/O) operations, whereas the former is running CPU-intensive operations.

The *port range* technique establishes a direct connection between processes that communicate with each other (Agullo et al., 2010). As a consequence, this approach has no overhead on the performance of the communication library: The bandwidth and the latency of the communications are the same as the those obtained by the Vanilla communication libraries.

Interconnecting processes through a proxy does not enable a direct connection between them. It introduces an extra "hop" between the two processes; hence, the process–process latency is the result of adding the two process–proxy latencies (Agullo et al., 2010). As a consequence, the physical location of the proxy with respect to the processes is of major importance in order to minimize this additional overhead. Besides, the proxy's bandwidth is shared between all the processes that are using it. If several processes are communicating at the same time through a given proxy, the available bandwidth for each process will be divided between them. Finally, performance evaluations have been conducted on "raw communications," and the impact of the performance obtained by these techniques on benchmark applications has been compared by Coti et al. (2008).

More comprehensive studies of QCG-OMPI and another grid-enabled application are presented by Agullo et al. (2010). A linear algebra factorization has been adapted to the grid using a communication-avoiding algorithm with grid-aware domain decomposition and a reduction algorithm that confines the communications within sets of processes that match the underlying physical topology. The performance of this application shows good scalability on the grid. Chapter 8 of this volume presents a topology-aware evolutionary algorithm and its application to gene regulatory network modeling and simulation.

9.3.2 QCG-ProActive

The vast number of Java-based legacy applications in use prompted developments attempting to provide a similar functionality for parallel Java applications as MPI offers to C/C++ or Fortran parallel code. Instead of exploiting existing Java bridges to MPI implementations, we decided to use the ProActive Parallel suite (Baduel et al., 2006). The library uses the standard Java RMI framework as a portable communication layer. With a reduced set of simple primitives, ProActive (version 3.9 as used in QCG) provides a comprehensive toolkit that simplifies the programming of applications distributed on local area networks, clusters, Internet grids, and peer-to-peer intranets for Java-based applications. However, when we designed QCG, the standard ProActive framework did not provide any support for multiuser environments, advance reservation, and cross-cluster coallocation. To meet the requirements of complex system simulation applications and users, we developed extensions to the ProActive library (called QCG-ProActive) with the following goals:

- To preserve standard ProActive library properties (i.e., allow legacy ProActive applications to be seamlessly ported to QCG)
- To provide end users with a consistent QCG-Broker (see further) Job Profile schema as a single document for describing application parameters required for execution as well as resource requirements (in particular, network topology and estimated execution time)
- To prevent end users from the necessity to have direct (i.e., over Secure Shell [SSH]) access to remote clusters and machines

9.3.2.1 *Cross-Cluster Deployment and Communication*   In the QCG environment, additional services were required in order to support the spawning of parallel application processes on coallocated computational resources. The main reason for this was that standard deployment methodologies used in OpenMPI and ProActive relied on either RSH/SSH or specific local queuing functionalities. Both are limited to single-cluster runs (e.g., the SSH-based deployment methods are problematic if at least one cluster has worker nodes that have private IP addresses). Those services are called the coordinators and are implemented as Web services. Taking into account different existing cluster configurations, we may distinguish the following general situations:

1. **A Computing Cluster with Public IP Addresses.** Both the front end and the worker nodes have public IP addresses. Typically, a firewall is used to restrict access to internal nodes.
2. **A Computing Cluster with Private IP Addresses.** Only the front-end machine is accessible from the Internet; all the worker nodes have private IP addresses. Typically, NAT is used to provide outbound connectivity.

These cluster configuration types influence intercluster communication techniques supported in QCG, called port range and *proxy* respectively.

9.3.2.2 *Port Range Technique*   The port range technique is a simple approach that makes the deployment of parallel environments firewall friendly. Most of the existing parallel environments use random ports by default to listen for incoming TCP/IP traffic. This makes cross-domain application execution almost impossible as most system administrators typically forbid to open all inbound ports to the Internet due to security reasons. By forcing the parallel

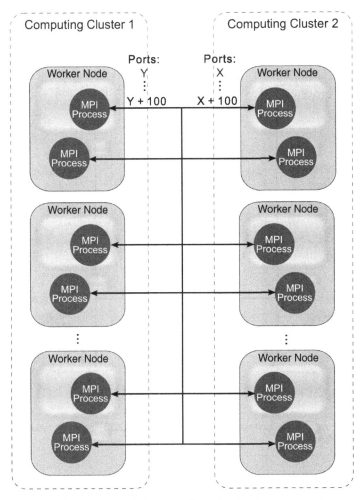

Figure 9.3 *A typical port range technique used by administrators of computational resources for a cross-cluster communication.*

environments to use only a predefined, unprivileged range of ports, it is much easier for administrators to configure the firewall in a way to allow incoming MPI and ProActive traffic without exposing critical system services to the Internet.

The port range technique is illustrated in Figure 9.3. Each of the site administrators has to choose a range of ports to be used (e.g., [5000–5100] for the parallel communication) and configure the firewall appropriately. One should note that the port range technique solves the problem of the cross-cluster connectivity for computing clusters where all worker nodes have public IP addresses.

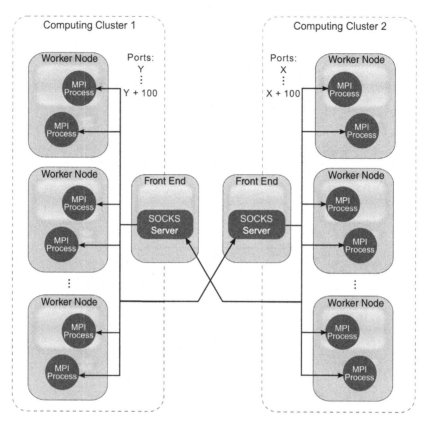

Figure 9.4 A new proxy technique for QCG-OMPI and QCG-ProActive based on SOCKS servers in QosCosGrid.

9.3.2.3 Proxy Technique In the second category of clusters where worker nodes have private IP addresses, the port range technique is not sufficient as all the worker nodes are not addressable from the outside networks. Therefore, in addition to the port range technique, which only shapes incoming traffic, we adopted a new proxy technique. In our approach, additional SOCKS proxy services have to be deployed on front-end machines to route incoming traffic to the MPI and ProActive processes running inside clusters on local worker nodes. The new proxy-based technique is depicted in Figure 9.4.

9.4 QCG MIDDLEWARE

In a nutshell, the QCG middleware consists of two logical levels: grid level and AD level. Grid-level services control, schedule, and generally supervise the execution of end-user applications, which are spread over independent

ADs. The AD represents a single resource provider (e.g., HPC or data center) participating in a certain grid or cloud environment by sharing its computational resources with both local and external end users. The logical separation of ADs corresponds to the fact that they are owned by different institutions or resource providers. Each institution contributes its resources for the benefit of the entire grid or cloud while controlling its own AD and own resource allocation/sharing policies. All involved organizations agree to connect their resource pools exposed by AD-level services to a trusted upper-level middleware, in this case, QCG middleware. Based on these assumptions, QCG middleware tries to achieve optimal resource utilization and to ensure the requested level of quality of service for all the end users. The key component of every AD in QCG is the QCG-Computing service, which provides remote access to queuing system resources. The QCG-Computing service supports advance reservation, parallel execution environments—OpenMPI and ProActive, with coordinators being responsible for the synchronization of cross-cluster executions—and data transfer services for managing input and output data. Another relevant service at the AD level is in charge of notification mechanisms: It is called QCG-Notification. All AD-level services are tightly integrated and connected to the grid-level services in QCG. There are two critical services at the grid-level: the QCG-Broker, which is a metascheduling framework controlling executions of applications on the top of queuing systems via QCG-Computing services, and the Grid Authorization Service (GAS), which offers dynamic, fine-grained access control and enforcement for shared computing services and resources. From an architectural perspective, GAS can also be treated as a trusted single logical point for defining security policies. The overall QCG architecture is depicted in Figure 9.5.

9.4.1 QCG-Computing Service

QCG-Computing (the successor of the SMOA Computing and OpenDSP projects) is an open architecture implementation of the SOAP Web service for multiuser access and policy-based job control routines by various queuing and batch systems managing local computational resources. This key service is using Distributed Resource Management Application API (DRMAA) to communicate with the underlying queuing systems (Troger et al., 2007). QCG-Computing has been designed to support a variety of plug-ins and modules for external communication as well as to handle a large number of concurrent requests from external clients and services, in particular, QCG-Broker and GAS. Consequently, it can be used and integrated with various authentication, authorization, and accounting services or to extend capabilities of existing e-infrastructures based on UNICORE, gLite, Globus Toolkit, and others. QCG-Computing service is compliant with the Open Grid Forum (OGF) HPC Basic Profile specification, which serves as a profile over OGF standards like Job Submission Description Language (JSDL) and Open Grid Services Architecture (OGSA) Basic Execution Service (OGF, 2007). In addition, it

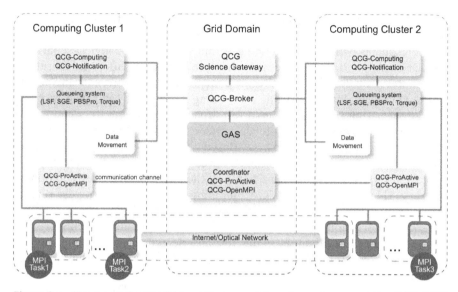

Figure 9.5   *The overall QosCosGrid architecture and its main services supporting QCG-OMPI and QCG-ProActive.*

offers remote interfaces for advance reservation management and supports basic file transfer mechanisms. QCG-Computing was successfully tested with the following queuing systems: Sun Grid Engine (SGE), Platform LSF, Torque/ Maui, PBS Pro, Condor, and Apple XGrid. Therefore, as a crucial component in QCG, it can be easily set up on the majority of computing clusters and supercomputers running the aforementioned queuing systems. Currently, advance reservation capabilities in QCG-Computing are exposed for SGE, Platform LSF, and Maui (a scheduler that is typically used in conjunction with Torque). Moreover, generic extensions for advance reservation have been proposed for the next DRMAA standard release.

9.4.2   QCG-Notification and Data Movement Services

QCG-Notification (the successor of SMOA Notification) is an open source implementation of the family of WS-Notification standards (version 1.3) (OASIS Standards, 2008). In the context of QCG, it is used to extend features provided by QCG-Computing by adding standards-based synchronous and asynchronous notification features. QCG-Notification supports the topic-based publish/subscribe pattern for asynchronous message exchange among Web services and other entities, in particular, services or clients that want to be integrated with QCG. The main part of QCG-Notification is based on a highly efficient, extended version of the Notification Broker, managing all items participating in notification events. Today, QCG-Notification offers sophisticated notification capabilities, for example, topic and message content notification

TABLE 9.1 Functional Comparison of Leading Notification Frameworks

	QCG-Notification	ServiceMix	WebSphere	GT 4.x
Language	Ansi C	Java	Java	C/Java
Type	Brokered	Brokered	Brokered	Base
Topic namespaces	Yes	No	Yes	Yes (flat only)
Dialect	Full	Simple	Full	Simple
Dynamic topics	Yes*)	Yes	Yes	Yes
Message filters	Yes	Yes	Yes	No
Pull points	Yes	Yes	Yes	No
Core functions	QCG-Core, XMPP	JBI services Bus JMS JMS	Enterprise Service bus	Java WS-Core C WS-Core

filtering and pull-and-push styles of transporting messages. QCG-Notification has been integrated with a number of communication protocols as well as various Web service security mechanisms. The modular architecture of QCG-Notification makes it relatively straightforward to develop new extensions and plug-ins to meet new requirements. In QCG, for example, it was used for brokering notification messages about the job state changes linking QCG-Broker and QCG-Computing. All instances of the QCG-Computing service act as information *producers*, while the QCG-Broker service is the *consumer* of job notifications in QCG. More sophisticated configurations of QCG-Notification with both QCG middleware services and external entities based on service-oriented architecture patterns are also possible. Table 9.1 presents the result of a functional comparison between QCG-Notification and other popular notification frameworks, namely, Apache ServiceMix (version 3.3.1), IBM WebSphere (version 7.0) and Globus Toolkit (version 4.2). A set of analyzed functional features covers most of the WS-Notification concepts, and it allows us to highlight fundamental differences among existing standards-compliant notification systems.

As many other e-infrastructures controlled by middleware services, QCG takes advantage of the GridFTP protocol for large data transfer operations, in particular, to stage in and stage out files for advanced simulations. GridFTP is a high-performance, secure, reliable data transfer protocol optimized for high-bandwidth wide area networks. It is a de facto standard for all data transfers in grid and cloud environments and extends the standard FTP protocol with functions such as third-party transfer, parallel and striped data transfer, self-tuning capabilities, X509 proxy certificate-based security, and support for reliable and restartable data transfers. The development of GridFTP is coordinated by the GridFTP Working Group under the hood of the OGF community.

9.4.3 QCG-Broker Service

QCG-Broker (formerly named GRMS) was designed to be an open source metascheduling framework that allows developers to build and easily deploy resource management systems to control large-scale distributed computing infrastructures running queuing or batch systems locally. Based on dynamic resource selection, advance reservation and various scheduling methodologies, combined with feedback control architecture, QCG-Broker deals efficiently with various metascheduling challenges, for example, coallocation, load balancing among clusters, remote job control, file staging support, or job migration (Kurowski et al., 2004). The main goal of QCG-Broker was to manage the whole process of remote job submission and advance reservation to various batch queuing systems and subsequently to underlying clusters and computational resources. It has been designed as an independent core component for resource management processes that can take advantage of various low-level core and grid services and existing technologies, such as QCG-Computing, QCG-Notification, or GAS, as well as various grid middleware services such as gLite, Globus, or UNICORE. Addressing various demanding computational needs of large-scale complex simulations, which in many cases can exceed capabilities of a single cluster, the QCG-Broker can flexibly distribute and control applications onto many computing clusters or supercomputers on behalf of end users. Moreover, owing to some built-in metascheduling procedures, it can optimize and run efficiently a wide range of applications while at the same time increasing the overall throughput of computing e-infrastructures. Advance reservation mechanisms are used to create, synchronize, and simultaneously manage the coallocation of computing resources located at different ADs. The XML-based job definition language Job Profile makes it relatively easy to specify the requirements of large-scale parallel applications together with the complex parallel communication topologies. Consequently, application developers and end users are able to run their experiments in parallel over multiple clusters.

Defined communication topologies may contain definitions of groups of MPI or ProActive processes with resource requirements, using resource and network attributes for internal and external group-to-group communication. Therefore, various application-specific topologies such as master–slave, all-to-all, or ring are supported in the Job Profile language. The Job Profile language has been adopted for complex system modeling and simulation purposes elsewhere (see, e.g., Chapter 8 of this volume or the work reported by Agullo et al., 2010).

To meet the requirements of complex scenarios consisting of many cooperating and possibly concurrent applications, for example, exchanging steering parameters in multiscale simulation studies, QCG-Broker is able to deal with complex applications defined as a set of tasks with precedence relationships (workflows). The workflow model built into QCG-Broker is based on direct acyclic graphs. In this approach, an end user specifies in advance precedence constraints of a task in the form of *task–state relationships*. What differentiates QCG-Broker from other middleware services supporting workflows is that every single task can be connected not only with input

TABLE 9.2 Comparison of QCG-Broker with Other Leading Brokering and Scheduling Services

Feature/System	QCG-Broker	Moab Grid Suites	CSF4	GUR	HARC
Negotiation protocol	Enhancement 1-phase commit	—	None	Reserve Cancel	Paxos
Economic support	Yes	No	No	No	No
Coallocation	Yes	Yes	No	Yes	Yes
Topology-aware coallocation	Yes	No	No	No	No
Scheduling on time axis	Yes	Yes	No	No	No
Local schedulers support	LSF Torque PBSPro SGE SLURM OpenPBS	Torque PBSPro LSF SGE SLURM OpenPBS LoadLeveler	LSF PBS Condor SGE	Catalina	LoadLeveler PBSPro LSF Torque
DRMAA support	Yes	Yes	No	No	No
Open source code	Yes	No	Apache	GPL	OpenSource
OpenDSP support	Yes	No	No	No	No

or output files but may also be triggered by predefined conditional rules or by the status of one or more jobs or tasks. Additionally, QCG-Broker supports parameter sweeps and allows to start in a single call multiple instances of the same application with different sets of arguments. For each task in the collection, the value of one or more of the task parameters may be changed in some predefined fashion, thus creating a parameter space. This is a very useful feature and gives the end user an easy way to search the parameter space for the concrete set of parameters that meet the defined criteria. QCG-Broker is a unique feature to allow end users to define multidimensional parameter spaces. Moreover, QCG-Broker has been successfully integrated with the Grid Scheduling SIMulator (GSSIM) to perform advanced metascheduling optimization, tuning, or reconfiguration experiments (Kurowski et al., 2007). Table 9.2 depicts the main features of QCG-Broker and compares these to other metascheduling frameworks deployed in e-infrastructures.

9.5 ADDITIONAL QCG TOOLS

9.5.1 Eclipse Parallel Tools Platform (PTP) for QCG

The PTP is intended to address a major deficiency in the development of parallel programs, namely, the lack of a robust open source targeted

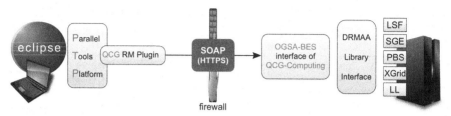

Figure 9.6   *Eclipse PTP plug-in for parallel application development and debugging integrated with QosCosGrid.*

development environment with tools that assist in software development for parallel applications. The PTP offers a variety of useful features for parallel application developers, in particular, a parallel integrated development environment, a scalable debugger, integration with parallel tools, and interaction with parallel systems. Currently, the main supported languages are C/C++ and Fortran. We used the Eclipse PTP framework to support QCG-OMPI and remote job submission, debugging, and monitoring using the QCG middleware, in particular, QCG-Computing services.

In order to connect to QCG-Computing services, it is necessary to provide connection data. These are inserted by the user via a graphical user interface (GUI) wizard page from the QCG RM plug-in. As the wizard is accepted, the data reach an object from class `QCGServiceProvider` responsible for keeping the data persistent. When the Eclipse PTP requests a connection name for the first time (which it is going to use for the connection), a new instance from the `QCGConnection` class is created and registered at `QCGConnectionManager`. This object contains the Web service ports for `rsync`, staging and activity management. Once the user requests to start the Resource Manager, classes `QCGResourceManager` and `QCGRuntimeSystem` are created and certain methods are called. These classes obtain basic data about the system (by calling `getFactoryAttributes()`) and display them in Eclipse's *parallel run-time* perspective. A generic architecture of a new Eclipse PTP plug-in available for QCG users and developers is depicted in Figure 9.6.

9.6   QOSCOSGRID SCIENCE GATEWAYS

The advanced Web-based graph- and multimedia-oriented user interfaces designed for scientists and engineers could change the way end users collaborate, deal with advanced simulations, share results, and work together to solve challenging problems. Moreover, future science and engineering gateways will influence the way end users will access not only their data but also control and monitor demanding computing simulations over the Internet. To allow end users to interact remotely with future supercomputers and large-scale computing environments in a more visual manner, we developed a Web tool called Vine Toolkit. Russell et al. (2008) demonstrated that this tool can be used as a core Web platform for various science gateways integrated with various e-infrastructures based on UNICORE, gLite, or Globus Toolkit middleware services. The Vine Toolkit is a modular, extensible, and easy-to-use tool as well as a high-level application program interface for various applications, visualization components, and building blocks to allow interoperability between a wide range of grid and supercomputing technologies. Similar to stand-alone GUIs, it supports Adobe Flex and BlazeDS technologies, allowing the creation of advanced Web applications. Additionally, the Vine Toolkit has been integrated with well-known open source Web frameworks, such as Liferay and GridSphere. Using the enhanced version of Vine Toolkit, we created a new Science Gateway called QCG Gateway. The QosCosGrid Science Gateway consists of a general part showing and monitoring computational resource characteristics as well as a set of domain-specific Web applications developed for certain complex system use cases. With these tools, end users are able to use only Web browsers to create and submit their complex simulations, monitor their progress, and access and analyze the results generated (Fig. 9.7).

The QCG e-infrastructure has been deployed on a large number of computational resources provided by various research centers, such as the Poznan Supercomputing and Networking Center (Poland), National Institute for Research in Computer Science and Control (INRIA) (France), and Dortmund University of Technology (Germany). A number of complex systems scientists

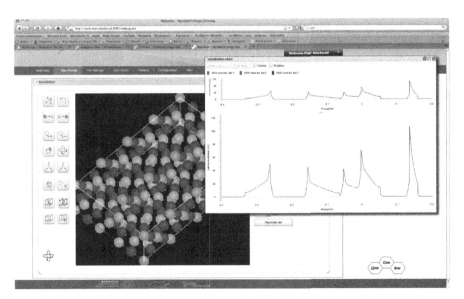

Figure 9.7 Example complex system simulations executed and controlled via the Web-based QosCosGrid Science Gateway.

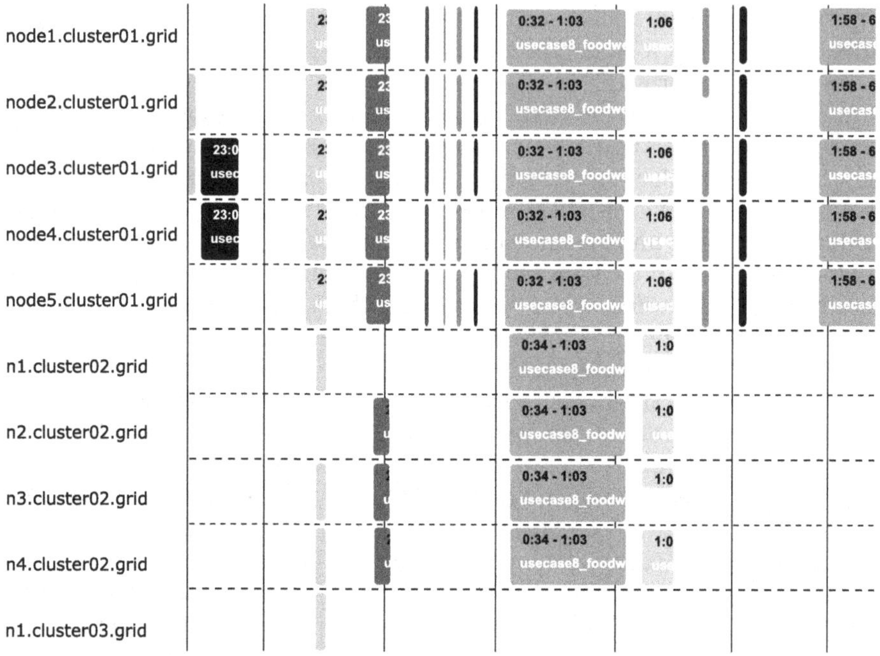

Figure 9.8 *Example coallocations using advance reservation QosCosGrid capabilities for cross-cluster parallel simulations.*

and modelers have performed various tests and benchmarks using QCG solutions. The QCG Gateway displays monitoring and resource information about the underlying infrastructure and provides a feedback not only for end users but also for resource owners and administrators (Fig. 9.8). Various monitoring tests are invoked periodically and their results are presented as graphical maps and diagrams in a Web browser. The first two maps show results of bidirectional tests of cross-domain QCG-OMPI and QCG-ProActive applications, measuring bandwidth and latency among front-end machines of computing clusters in different ADs connected via the Internet. We have also added useful Gantt charts displaying local and cross-domain job execution information as well as information on advance reservation and coallocation of computing resources. Both the test-bed monitoring and GUI layers were developed using Adobe Flex-based components.

9.7 DISCUSSION AND RELATED WORK

Existing grid e-infrastructures, such as TeraGrid in the United States and CERN's Enabling Grids for E-sciencE (EGEE)/LCG in Switzerland, Grid'5000 in France, D-Grid in Germany, or PL-GRID in Poland, provide thousands of computational resources, offering facilities similar to large-scale parallel computing production environments. One of the critical components for the execution of advanced applications in such environments is the *metascheduler*. Its main role is to schedule complex applications that end users wish to run on available computational resources and to guarantee "topology awareness." Topology awareness implies that matching of the resource offers and requests must take into account not only the computational characteristics of the resources but also their interconnections. To the best of our knowledge, these kinds of requirements have not been addressed by any of the existing metaschedulers. Thus, the majority of the up-to-date computing e-infrastructures support relatively easy metascheduling strategies, in particular, easy matching-based techniques, without advance reservation and coallocation of computational resources. Typical systems include the EGEE Workload Management System, eNANOS, and GridWay. The relatively simple metascheduling approach adopted by these systems is sufficient to execute "embarrassingly parallel" applications, such as parameter sweeps or MPI applications, within a *single* computing cluster. This approach limits end users to the computing power of one computing cluster, which may not be sufficient to perform more sophisticated complex system simulations. For cross-cluster MPI and ProActive applications, the requested and offered network topology cannot be ignored, meaning that topology-aware scheduling, together with coallocation capabilities, must be employed. These features have been successfully implemented and integrated to provide a new QCG e-infrastructure for complex system modelers, scientists, and end users.

Emerging e-infrastructures offer new quality-of-service capabilities as well as new hybrid programming and execution environments such as MPI/OpenMP and MPI/CUDA parallel hybrids. Taking advantage of the results presented in this chapter, the adaptation of QCG solutions to many-core systems would be an interesting area of our future research. Another natural direction of future research and development would be to explore new optimization criteria, for example, energy consumption or heating, as these have become critical in large distributed computing environments. As virtualization techniques of operating systems and embedded applications are maturing, another fruitful avenue of future research revolves around efficient resource management techniques using the virtual machine concept. Moreover, the recent technological shift from the area of multicore to many-core systems and systems-on-the-chip introduced another scheduling layer at the operating system level. Future large-scale computing systems will have to deal with a hierarchy of complex static and reconfigurable computing structures, and new resource management techniques need to be invented to control such systems on a very low level of granularity.

Ethernet technology has been the dominant data link protocol in local area networks for many years, and it is widely offered as a customer service, but it was not designed for large carrier-scale transport networks. However, because of its suitability for data and multimedia applications, its flexibility, and its widespread use, many carriers now consider Ethernet as a potential convergence solution for next-generation networks. Research grid networks link distributed computing resources to perform highly demanding computations needing vast processing capability. To make optimum use of the computing resources, high-capacity, low-latency connections are needed, and complex optimization algorithms need to calculate the optimal exploitation of the computing resources, taking into account factors such as the data rate offered by the connecting links and their latency.

Currently, QCG is extensively tested by its developers and by external research communities interested in new computing e-infrastructures for distributed multiscale simulations across disciplines. Driven by seven challenging applications from five representative scientific domains (fusion, clinical decision making, systems biology, nanoscience, and engineering), they will deploy a computational science environment for distributed multiscale computing under the MAPPER[4] project based on QCG technologies. Various extensions proposed to QCG will result in high-quality components for today's e-infrastructures in Europe by enabling distributed execution of two modes (loosely and tightly coupled) of multiscale computing in a user-friendly and transparent way.

REFERENCES

E. Agullo, C. Coti, J. Dongarra, et al. QR factorization of tall and skinny matrices in a grid computing environment. In *Proc. of the 24th IEEE Int'l Parallel and Distributed Processing Symposium*, 2010.

E. Agullo, C. Coti, T. Herault, et al. QCG-OMPI: MPI applications on grids. *Future Generation Computer Systems*, 27(4):357–369, 2011.

L. Baduel, G. Baude, D. Caromel, et al. Programming, deploying, composing, for the grid. In J. C. Cunha and O. F. Rana, editors, *Grid Computing: Software Environments and Tools*, pp. 205–229, London: Springer-Verlag, 2006.

S. T. Barnard. PMRSB: Parallel multilevel recursive spectral bisection. In *Proc. of the 1995 ACM/IEEE Conference on Supercomputing, Supercomputing '95*, New York: ACM, 1995.

C. Coti, T. Herault, S. Peyronnet, et al. Grid services for MPI. In ACM/IEEE, editor, *Proc. of the 8th IEEE Int'l Symposium on Cluster Computing and the Grid (CCGrid'08)*, pp. 417–424, Lyon, France: IEEE Computer Society, 2008.

P.-O. Fjallstrom. Algorithms for graph partitioning: A survey. *Linkoping Electronic Articles in Computer and Information Science*, 3(10):1–34, 1998.

L. Gulyás, G. Szemes, G. Kampis, et al. A modeler-friendly API for ABM partitioning. In *Proc. of the ASME 2009 Conference*, San Diego, California, USA, 2008.

K. Kurowski, W. de Back, W. Dubitzky, et al. *Complex System Simulations with QosCosGrid*, Vol. 5544, pp. 387–396, Berlin and Heidelberg: Springer, 2009.

---

[4] http://www.mapper-project.eu.

K. Kurowski, B. Ludwiczak, J. Nabrzyski, et al. Dynamic grid scheduling with job migration and rescheduling in the GridLab resource management system. *Scientific Programming*, 12:263–273, 2004.

K. Kurowski, J. Nabrzyski, A. Oleksiak, et al. Grid scheduling simulations with GSSIM. In *Proc. of the 13th Int'l Conference on Parallel and Distributed Systems—Vol. 2*, pp. 1–8, Washington, DC: IEEE Computer Society, 2007.

K. Kurowski, T. Piontek, P. Kopta, et al. Parallel large-scale simulations in the pl-grid environment. In M. Stroinski et al., editors, *Computational Methods in Science and Technology*, pp. 47–56, Poznan, Poland, 2010.

M. Mazzariol, B. A. Gennart, and R. D. Hersch. Dynamic load balancing of parallel cellular automata, 2000.

OASIS Standards. OASIS web services notification technical committee, 2008. URL http://www.oasis-open.org.

OGF. Open grid forum, 2007. http://ogf.org.

M. Russell, P. Dziubecki, P. Grabowski, et al. The Vine Toolkit: A Java framework for developing grid applications. In R. Wyrzykowski et al., editors, *Parallel Processing and Applied Mathematics, volume 4967 of Lecture Notes in Computer Science*, pp. 331–340, Berlin and Heidelberg: Springer, 2008.

M. Scheutz and P. Schermerhorn. Adaptive algorithms for the dynamic distribution and parallel execution of agent-based models. *Journal of Parallel Distributed Computing*, 66:1037–1051, 2006.

P. M. A. Sloot and A. G. Hoekstra. Multi-scale modelling in computational biomedicine. *Briefings in Bioinformatics*, 11(1):142–152, 2010.

P. Troger, H. Rajic, A. Haas, et al. Standardization of an API for distributed resource management systems. In *Proc. of the 7th IEEE Int'l Symposium on Cluster Computing and the Grid, CCGRID'07*, pp. 619–626, Washington, DC: IEEE Computer Society, 2007.

Glossary

ABMS See *agent-based modeling and simulation.*

Agent The individually identifiable component of an agent-based model. An agent typical executes some type of behavior and represents a decision maker at some level.

Agent-based modeling and simulation (ABMS) A method of simulation that computes the system-level consequences of the behaviors of individuals. ABMS is also known as *individual-based simulation* or *individual-based modeling.*

Ahmad–Cohen neighbor scheme (ACS) Division of gravitational force acting on a particle into two components, an irregular one (from nearby particles) and a regular one, from distant particles. Direct N-body codes of Aarseth of NBODY5 and higher (NBODY6, NBODY6++) employ ACS. Each part of the force can be used to define a separate time step, which determines the update interval for this force component only, while extrapolating the other component with low-order Taylor series polynomials, if needed. The idea of the ACS could be generalized to more levels, but this seems not to be in common use at this time.

Base Object Model (BOM) A Simulation Interoperability Standards Organization (SISO) standard for describing a reusable piece part of a simulation.

BOM See *Base Object Model.*

Large-Scale Computing, First Edition. Edited by Werner Dubitzky, Krzysztof Kurowski, Bernhard Schott.
© 2012 John Wiley & Sons, Inc. Published 2012 by John Wiley & Sons, Inc.

CCA See *Common Component Architecture.*

Cloud computing Cloud computing describes a computing model for IT services based on the Internet. The fundamental concept of cloud computing is that processing and data do not reside in a specified, known, or static location. Typically, cloud computing infrastructures provide dynamically scalable, virtualized resources over the Internet. The user perspective of cloud computing takes the form of Web-based tools and applications that are accessed via a Web browser.

Cluster See *cluster computer.*

Cluster computer A collection of linked computers working closely together to "emulate" a single computer powerful computer. The components of a cluster computer are typically connected through fast local area networks. Clusters are usually deployed to improve performance and availability while being much more cost-effective than single computers of comparable specification.

Common Component Architecture (CCA) A standard for component-based software engineering used in high-performance (also known as scientific) computing.

Computational model Refers to the computer realization (implemented software) of a *simulation model.*

Discrete-event simulation A simulation technique where the operation of a system is modeled as a series of events. These events are ordered internally with respect to one another (i.e., earlier events occur before later ones) such that the simulation proceeds by removing events from a queue and executing them.

Distributed simulation system An application consisting of distributed simulation modules.

e-infrastructure An electronic computing and communication environment (hardware, software, data and information repositories, services and digital libraries, and communication networks and protocols) and the people and organizational structures needed to support large-scale research and development efforts. e-Infrastructures enable collaborations across geographically dispersed locations by providing shared access to unique or distributed facilities, resources, instruments, services, and communications. e-Infrastructures usually consist of (1) high-performance communication networks connecting the collaborating sites; (2) grid computing to facilitate the sharing of nontrivial amounts of resources such as processing elements, storage capacity, and network bandwidth; (3) supercomputing capabilities to address large-scale computing tasks; (4) databases and information repositories that are shared by the participating organizations; (5) globally operating research and development communities that collaborate to solve highly challenging scientific and engineering problems; and (6) standards to enable the sharing, interoperation, and communication in e-infrastructures.

Evolutionary algorithm A generic population-based optimization algorithm based on iteration and the principles of evolution (variation, selection, and heredity). At each iteration, operators create diversity and a fitness function evaluates individuals within the population based on some characteristics. Individuals correspond to potential solutions and the fittest individual corresponds to the most optimal solution.

Federation object model (FOM) Describes the shared objects, attributes, and interactions for a distributed simulation system based on HLA.

FOM See *federation object model.*

Gene regulatory network A set of genes that interact with each other via the products they express, namely, forms of RNA and proteins. Interacting genes may activate or *switch on* the expression of another gene, or they may repress or *switch off* the expression of another gene. Gene regulatory networks are used by cells to control their life-support mechanisms.

GPU See *graphical processing unit.*

Graphical processing unit (GPU) A specialized microprocessor that handles and accelerates 3-D or 2-D graphic rendering. Owing to their highly parallel structure, GPUs are very effective not only in graphics computing but also in a wide range of other complex computing tasks.

GRAPE See *Gravity Pipe.*

Gravitational potential softening Softening of the singularity of Newtonian (or Coulombian) potential Φ between particles. In astrophysical N-body simulations, a Plummer softening is often used, which is $\Phi(r) = \Phi_0 \big/ \sqrt{r^2 + \varepsilon^2}$ for the potential of a single particle (r is the distance from the particle's center; ε is the scaling radius for the size of the particle). It is the true potential of a gas sphere whose density is given by Plummer's model, which is in fact an $n = 5$ polytrope. Note that Plummer softening differs from the true Newtonian potential $\Phi(r) = \Phi_0/r$ at all radii. More modern variants of softening, which are used for smoothed particle hydrodynamics, use softening kernels, which differ from the Newtonian potential only on a compact subspace around the particle (of the order of a few ε).

Gravity Pipe (GRAPE) A special-purpose computer designed to accelerate the calculation of forces in simulations of interacting particles. GRAPE systems have been used for N-body calculations in astrophysics, molecular dynamics models, the study of magnetism, and many other applications.

Grid computing Grid computing combines computer resources from multiple administrative domains to run complex computing applications. Typically, grid computing environments are less tightly coupled, more heterogeneous, and more geographically dispersed than conventional cluster and supercomputers.

HAND Highly Available Dynamic Deployment Infrastructure for Globus Toolkit facilitates dynamic deployment of grid services.

Hermite scheme Two-point interpolation scheme that uses a function value and its a priori known first derivative in order to obtain fourth-order accuracy. In gravitational N-body simulations, the gravitational force and its time derivative are used to have a fourth-order time integration scheme, with the need to store only particle data for two points in time, which is convenient for parallelization and memory management. Recently, generalizations have been described going up to eighth-order time integration, which requires then a priori knowledge of up to the fourth derivative of the gravitational force.

High-level architecture (HLA) An IEEE standard for distributed computer simulation systems. Communication between simulations is managed by a **run-time infrastructure** (RTI).

High-performance computing (HPC) HPC uses supercomputers, computer clusters, or other large-scale or distributed computing technology to solve large-scale computing problems. Nowadays, any computer system that is capable of a teraflop computing performance is considered an HPC computer.

HLA See *high-level architecture*.

HPC See *high-performance computing*.

Job Usually, a grid *job* is a binary executable or command to be run in a remote resource (machine). The remote server is sometimes referred to as a "contact" or a "gatekeeper." When a job is submitted to a remote gatekeeper (server) for execution, it can run in two different modes: batch and nonbatch. Usually, when a job runs in batch mode, the remote submission call will return immediately with a job identifier, which can later be used to obtain the output of the call. In the nonbatch job submission mode, the client will wait for the remote gatekeeper to follow through with the execution and will return the output. Batch mode submission is useful for jobs that take a long time, such as process-intensive computations.

Master–worker pattern The master–worker pattern is used in distributed computing to address easy-to-parallelize large-scale computing tasks. This pattern typically consist of two types of entities: *master* and *worker*. The master initiates and controls the computing process by creating a *work set* of tasks and putting them in some "shared space" and then waits for the tasks to be pulled from the space and completed by the workers. One of the advantages of the master–worker pattern is that it automatically balances the load because the work set is shared and the workers continue to pull work from the set until there is no more work to be done. Algorithms implementing the master–worker pattern usually scale well, provided that the number of tasks is much higher than the number of workers and that the tasks are similar in terms of the amount of time they need to be completed.

MCT See *Model Coupling Toolkit*.

Message Passing Interface (MPI) A message passing parallel programming model in which data are moved from the address space of one process to that of another process through cooperative operations on each process. The MPI standard includes a language-independent message passing library interface specification.

Model Coupling Toolkit (MCT) A set of tools for coupling message passing parallel models.

Model reduction Refers to the approximation of a model aiming at a simplified model that is easier to analyze but preserves the essential properties of the original model.

MPI See *Message Passing Interface*.

Multiscale Coupling Library and Environment (MUSCLE) A software framework for building simulations according to the complex automata theory.

Multiscale Multiphysics Scientific Environment (MUSE) A software environment for astrophysical applications in which different simulation models of star systems are incorporated into a single framework.

MUSCLE See *Multiscale Coupling Library and Environment*.

MUSE See *Multiscale Multiphysics Scientific Environment*.

Object-oriented programming (OOP) A programming paradigm in which data and the operations on that data are encapsulated in an *object*. Other features of OOP include inheritance, where one object may inherit the data and operations of *parent object(s)*, and polymorphism, where an object can be used in place of its parent object while retaining its own behavior.

Ordinary differential equation In chemical kinetic theory, the interactions between species are commonly expressed using ordinary differential equations (**ODE**s). An ODE is a relation that contains *functions* of only one independent variable (typically t) and one or more of its derivatives with respect to that variable. The order of an ODE is determined by the highest derivative it contains (e.g., a first-order ODE involves only the first derivative of the function). The equation $5x(t) + \dot{x}(t) = 17$ is an example of a first-order ODE involving the independent variable t, a function of this variable, $x(t)$, and a derivative of this function, $\dot{x}(t)$. Since a derivative specifies a rate of change, such an equation states how a function changes but does not specify the function itself. Given sufficient initial conditions, various methods are available to determine the unknown function. The difference between ordinary differential equations and partial differential equations is that partial differential equations involve partial derivatives of several variables.

Parallel discrete-event simulation A distributed version of *discrete-event simulation* in which the events and their execution may occur in parallel across machines or processes. This usually includes some mechanism(s) for synchronizing the execution of events across machines or processes.

Parameter sweep A technique to explore or characterize a process, procedure, or function by means of a carefully generated set of input parameter combinations or configurations. The term *parameter* may cover a wide range of concepts, including structured data files, numeric or symbolic values, vectors or matrices, or executable models or programs. For example, a parameter sweep experiment may generate suitable inputs to explore a cost function or to create the energy surface for a 3-D graph. Sensitivity analysis could be viewed as a form of parameter sweep experiment where inputs are systematically varied to analyze how sensitive a model responds to variations of individual or groups of state variables. In contrast to *parameter estimation* and *parameter optimization* techniques, the parameter sweep procedure does not normally incorporate feedback from the output of a process or model to iteratively steer and adapt the generation of parameter combinations.

Partial differential equation Similar to an *ordinary differential equation* except that it involves functions with more than one independent variable.

Regularization Classical method of celestial mechanics to transform equations of motions for the two-body problem to a harmonic oscillator problem, removing the coordinate singularity at zero separation and allowing analytic continuation of solutions through the zero separation point. Regularization is commonly used for some direct N-body simulation codes to allow a more efficient integration of perturbed dominant pairs of particles at nonzero separation.

RTI See *run-time infrastructure*.

Run-time Infrastructure (RTI) A communication bus used by HLA simulation modules.

SAN See *storage area network*.

SCA See *Service Component Architecture*.

Sensitivity analysis An important tool to study the dependence of systems on their parameters. Sensitivity analysis helps to identify those parameters that have a significant impact on the system output and to capture the essential characteristics of the system. Sensitivity analysis is particularly useful for complex biological networks with a large number of variables and parameters.

Service A network-enabled entity that provides a specific capability, for example, the ability to move files, create processes, or verify access rights. A service is defined in terms of the protocol one uses to interact with it and the behavior expected in response to various protocol message exchanges (i.e., *service = protocol + behavior.*). A service may or may not be persistent (i.e., always be available); be able to detect and/or recover from certain errors; run with privileges; or have a distributed implementation for enhanced scalability. If variants are possible, then discovery mechanisms that allow a client to determine the properties of a particular instantiation of a service are important.

Service Component Architecture (SCA)   A set of specifications that describe a model for building applications and systems using a service-oriented architecture.

Service-oriented architecture (SOA)   An SOA is a collection of services that communicate with each other by simple data passing or to coordinate the activity of multiple services. Some means of connecting services to each other is needed. The main advantages of SOA include loose coupling, ease and flexibility of reuse, scalability, interoperability, and service abstraction from underlying technology.

Simulation Interoperability Standards Organization (SISO)   An international organization dedicated to the promotion of modeling and simulation interoperability and reuse for the benefit of a broad range of modeling and simulation communities

Simulation model   A formal (mathematical) description of a natural phenomenon or an engineering artifact used to simulate the behavior of the phenomenon or artifact. To facilitate efficient simulation that a simulation model is typically implemented as a computer program or software (referred to as *computational model*).

SISO   See *Simulation Interoperability Standards Organization.*

SOA   See *service-oriented architecture.*

Storage area network (SAN)   A storage area network is a special type of network that is separate from LANs and WANs. A SAN is usually dedicated to connect all the storage resources connected to various servers.

Supercomputer   A supercomputer is a computer that is at the front of current processing capacity. Supercomputers are typically used for highly calculation-intensive tasks such as problems involving quantum physics, weather forecasting, climate research, molecular modeling, physical simulations, and so on. In the context of supercomputing, the distinction between *capability computing* and *capacity computing* is becoming relevant. Capability is normally concerned with maximum computing power to solve a large problem in the shortest amount of time. Capacity computing, on the other hand, is concerned with efficient, cost-effective computing power to solve somewhat large problems or many small problems or to prepare for a run on a capability system.

Virtualization   In computing, virtualization refers to the creation of a virtual (rather than actual) version of something such as an operating system (OS), a server, a storage device, or network resources. The usual goal of virtualization is to centralize administrative tasks while improving scalability and workloads. Virtualization is part of a trend in which computer processing power is seen as a utility that clients can pay for only as needed. Some common virtualizations include the following: *Hardware virtualization* refers to the execution of software in an environment separated from the underlying hardware resources. *Memory virtualization* refers to the aggregation RAM resources from networked systems into a single memory pool. *Storage virtualization* refers to the separation of logical storage from physical storage. *Data virtualization* refers to the presentation of data as an abstract layer, independent of underlying database systems, structures, and storage. *Database virtualization* refers to the decoupling of the database layer, which lies between the storage and application layers within the application stack. *Network virtualization* refers to the creation of a virtualized network addressing space within or across network subnets. *Software virtualization*: (1) *OS-level virtualization* refers to the hosting of multiple virtualized environments within a single OS instance; (2) *application virtualization* refers to the hosting of individual applications in an environment separated from the underlying OS; and (3) *virtual machine* refers to a software implementation of a computer that executes programs like a real computer.

Virtual machine   A software implementation of a computer that executes programs like a physical machine. Software running inside a virtual machine is limited to the resources and abstractions provided by the virtual machine—it cannot cross the boundaries of the "virtual world" established by the virtual machine. Virtual machines are categorized into two major types: (1) a *system virtual machine*, which provides a complete system platform that supports the execution of a complete operating system, and (2) a *process virtual machine*, which is designed to run a single program, which means that it supports a single process.

Virtual organization (VO)   Refers to a set of individuals and/or institutions that are related to one another by some level of trust and rules of sharing resources, services, and applications.

VO   See *virtual organization.*

Web service (WS)   Refers to a Web-based application that uses open, XML-based standards and transport protocols to exchange data with clients. A WS is designed to support interoperable machine-to-machine interaction over a network.

Web Service Resource Framework (WSRF)   A Web service extension providing a set of operations that Web services may implement to become stateful.

WS   See *web service.*

WSRF   See *Web Service Resource Framework.*

Index

WILEY SERIES ON PARALLEL AND DISTRIBUTED COMPUTING
Series Editor: Albert Y. Zomaya

Printed in the United States
By Bookmasters